I0516515

Tucholsky Wagner Zola Scott Sydow Schlegel
Turgenev Wallace Fonatne Freud
Twain Walther von der Vogelweide Fouqué Friedrich II. von Preußen
Weber Freiligrath Frey
Kant Ernst
Fechner Fichte Weiße Rose von Fallersleben Richthofen Frommel
Engels Fielding Hölderlin
Fehrs Faber Flaubert Eichendorff Tacitus Dumas
Eliasberg Ebner Eschenbach
Feuerbach Maximilian I. von Habsburg Fock Eliot Zweig
Ewald Vergil
Goethe London
Mendelssohn Balzac Shakespeare Elisabeth von Österreich Ganghofer
Lichtenberg Rathenau Dostojewski
Trackl Stevenson Doyle Gjellerup
Mommsen Tolstoi Lenz Hambruch
Thoma Hanrieder Droste-Hülshoff
Dach Verne von Arnim Hägele Hauff Humboldt
Karrillon Reuter Rousseau Hagen Hauptmann Gautier
Garschin Defoe Baudelaire
Damaschke Descartes Hebbel
Hegel Kussmaul Herder
Wolfram von Eschenbach Dickens Schopenhauer Rilke George
Bronner Darwin Melville Grimm Jerome Bebel Proust
Campe Horváth Aristoteles Federer
Bismarck Vigny Barlach Voltaire Herodot
Gengenbach Heine
Storm Casanova Tersteegen Grillparzer Georgy
Chamberlain Lessing Langbein Gilm Gryphius
Brentano Claudius Schiller Lafontaine
Strachwitz Bellamy Schilling Kralik Iffland Sokrates
Katharina II. von Rußland Gerstäcker Raabe Gibbon Tschechow
Löns Hesse Hoffmann Gogol Wilde Gleim Vulpius
Luther Heym Hofmannsthal Klee Hölty Morgenstern
Roth Heyse Klopstock Kleist Goedicke
Luxemburg Puschkin Homer Mörike Musil
La Roche Horaz
Machiavelli Kierkegaard Kraft Kraus
Navarra Aurel Musset Lamprecht Kind Kirchhoff Hugo Moltke
Nestroy Marie de France Laotse Ipsen Liebknecht
Nietzsche Nansen Ringelnatz
Marx Lassalle Gorki Klett Leibniz
von Ossietzky May vom Stein Lawrence Irving
Petalozzi Platon Pückler Michelangelo Knigge Kafka
Sachs Poe Liebermann Kock
Korolenko
de Sade Praetorius Mistral Zetkin

The publishing house tredition has created the series **TREDITION CLASSICS**. It contains classical literature works from over two thousand years. Most of these titles have been out of print and off the bookstore shelves for decades.

The book series is intended to preserve the cultural legacy and to promote the timeless works of classical literature. As a reader of a **TREDITION CLASSICS** book, the reader supports the mission to save many of the amazing works of world literature from oblivion.

The symbol of **TREDITION CLASSICS** is Johannes Gutenberg (1400 – 1468), the inventor of movable type printing.

With the series, tredition intends to make thousands of international literature classics available in printed format again – worldwide.

All books are available at book retailers worldwide in paperback and in hardcover. For more information please visit: www.tredition.com

tredition was established in 2006 by Sandra Latusseck and Soenke Schulz. Based in Hamburg, Germany, tredition offers publishing solutions to authors and publishing houses, combined with worldwide distribution of printed and digital book content. tredition is uniquely positioned to enable authors and publishing houses to create books on their own terms and without conventional manufacturing risks.

For more information please visit: www.tredition.com

Theory of Silk Weaving A Treatise on the Construction and Application of Weaves, and the Decomposition and Calculation of Broad and Narrow, Plain, Novelty and Jacquard Silk Fabrics

Arnold Wolfensberger

Imprint

This book is part of the TREDITION CLASSICS series.

Author: Arnold Wolfensberger
Cover design: toepferschumann, Berlin (Germany)

Publisher: tradition GmbH, Hamburg (Germany)
ISBN: 978-3-8491-8775-0

www.tredition.com
www.tredition.de

Copyright:
The content of this book is sourced from the public domain.

The intention of the TREDITION CLASSICS series is to make world literature in the public domain available in printed format. Literary enthusiasts and organizations worldwide have scanned and digitally edited the original texts. tredition has subsequently formatted and redesigned the content into a modern reading layout. Therefore, we cannot guarantee the exact reproduction of the original format of a particular historic edition. Please also note that no modifications have been made to the spelling, therefore it may differ from the orthography used today.

PREFACE

The silk industry of America has of late years rapidly advanced to the front rank among the great textile industries of the world. It may indeed be proud of this position, to which that enterprising spirit and untiring energy peculiar to our nation, combined with our great technical and natural resources, has brought it.

That we are, on the other hand, not yet at the height of perfection we are also compelled to acknowledge, but if we consider the short space of time that the American industry has required for its development, as compared to the decades, almost centuries, to which some of the great European silk centers can look back, the fact is neither surprising nor discouraging.

While it must not be our aim to imitate or copy their ways, inasmuch as out conditions and circumstances are quite different from theirs, we may still profitably study their methods in order to overcome our deficiencies.

The greatest advantage which our competitors derive from such a long existence consists in having at their disposal a force of skilful, trained help. The manufacturers, appreciating the importance of this factor, make great efforts and pecuniary sacrifices to elevate and maintain the high standard of their industry.

For instance, they support textile schools and lecture courses, where young men can acquire a thorough technical education and equip themselves for a career of usefulness, thereby serving their own interests and at the same time furthering those of their chosen profession.

[pg 6]

This beneficial influence cannot fail to exert itself from the standard of the higher employer down to that of the weaver, who would naturally take more pains and interest in his work than if he were a mere mechanical appendage to his loom in order to keep it in motion.

Very little has been done in his country for technical education as far as the silk industry is concerned, and it was on this special branch, that prompted the author to offer in the present little work a treatise on the theory of shaft weaving for broad silks and ribbons.

It is divided into three principal parts:

1st. Drawing-in the warp in the harness.

2nd. The weaves and their application.

3rd. Decomposition or analysis of the cloth.

To the foregoing there have been added in the revised and enlarged edition several additional parts covering the following: JACQUARD WEAVES, BOX LOOM WEAVES, including CREPES, and COST CALCULATIONS for plain and fancy weaves.

The subject while condensed, is made as clear and comprehensible as possible, and to many desirous of increasing their knowledge in this direction, this should prove a valuable help.

The author, through the medium of this work, hopes to win the approval and encouragement of the manufacturers, and will feel amply repaid should his efforts tend to develop a deeper interest in the "Queen of Textiles."

[pg 7]

THEORY OF SILK WEAVING

DRAWING-IN

With this term we designate the operation preceding the weaving, by which all the warp-threads are drawn through the heddles of the harness.

The order in which this is done varies according to the weave and the nature of the fabric to be produced; so we distinguish:

Straight draws,

Skip draws,

Point draws,

Section draws.

STRAIGHT DRAWS

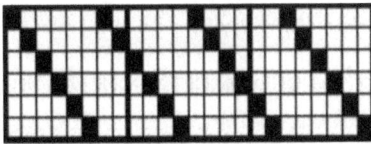

Fig. 1

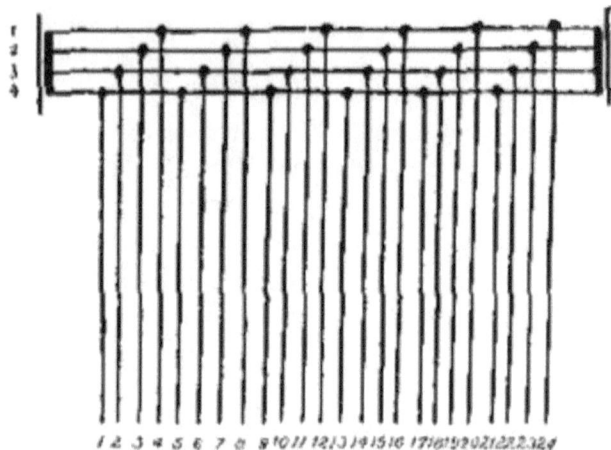

Fig. 2

These form the simplest and most common method of drawing-in. We begin with the first heddle on the left side of the shaft *nearest to the warp-beam*, then take the first heddle of second shaft and so on until all the shafts the set contains are used in rotation. This completes one "draw," and this operation is repeated until all the warp-threads are taken up.

The method of making the shaft nearest to the warp-beam the first, is almost universal with the silk business and is technically called *drawing-in from back to front*.

The opposite, or drawing in from *front to rear*, is used occasionally, however, and in this case makes the first heddle on the left hand side of the front shaft No. 1.

The making out of the *Drawing-in Draft*, which must indicate the arrangement or the rotation in which the warp-threads are drawn in, can be done in various ways, of which we will mention the two most popular methods. The first is by using common designing paper, and indicating the rotation by dots. The horizonal rows of squares represent the shafts, the vertical rows the warp-threads. Fig. 1 shows four repeats of a straight draw on six harness marked out according to this idea. A second method is to use paper ruled horizontally, the lines representing the shafts; and to draw vertical lines for the warp-threads. The latter are made to stop on the lines bearing the number of the shafts into which the respective threads are to be drawn. Fig. 2 is such a draft, illustrating six repeats of a draw on four harness from "Front to Rear."

SKIP DRAWS

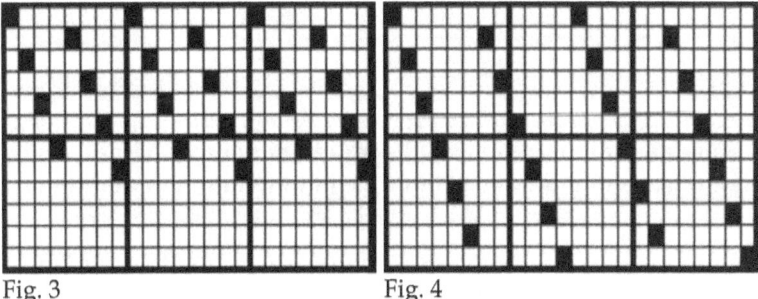

Fig. 3   Fig. 4

The draws coming under this heading are used very extensively in silk weaving, especially for fabrics requiring a heavy warp and a large number of shafts. Enter first the odd and then the even shafts. An 8 harness draw of this kind, of which three repeats are shown in Fig. 3, runs as follows: 1, 3, 5, 7, 2, 4, 6, 8.

Fig. 4 is a 12 harness draw of the same class.

POINT DRAWS

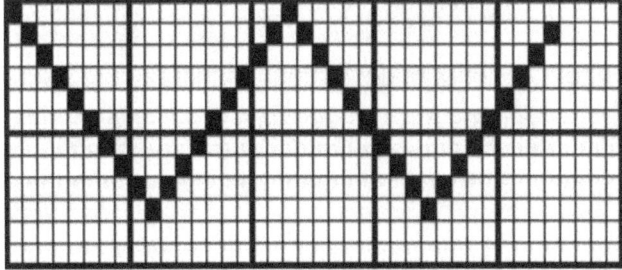

Fig. 5

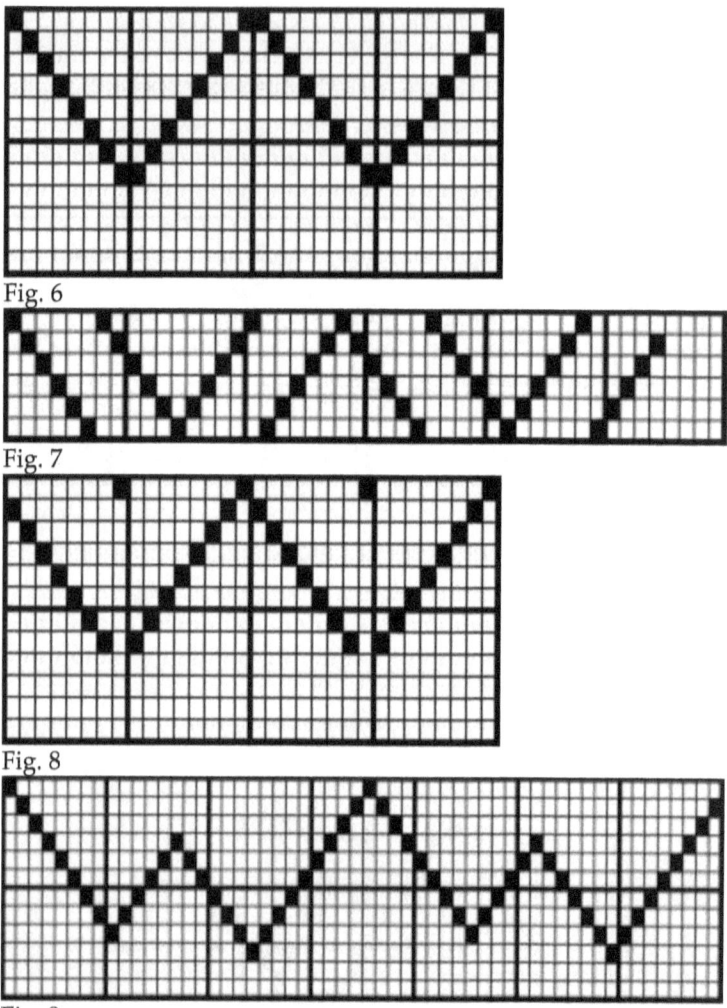

Fig. 6

Fig. 7

Fig. 8

Fig. 9
[pg 11]

Point draws are a combination of a regular straight draw from back to front and one from front to back, the first and the last shafts only being used once, while the rest receive two ends each in one repeat of the draw. Fig. 5 illustrates a regular point draw in 2 repeats on 10 shafts. It will be seen that 14 ends make a repeat; in fact, the number of warp-threads required for one draw will always be double the number of harness less 2, hence a 12 harness regular point draw will require 22 warp-threads for a repeat.

The drawing-in draft illustrated in Fig. 6 is a slight variation of the regular point draw; it consists, as will be seen, of a draw from back to front, and also a full one from front to back, there by causing a *double point*.

Another change from the regular point draw is illustrated in Figs. 7 and 8; this class may be called *Broken point draws*, because a new draw is begun before the other one is complete. Fig. 9 also comes in this class and represents a *zigzag draw* on 10 harness.

The drawing-in drafts which we have described under the head of "Point draws," are used mostly to obtain the various pointed and zigzag effects.

SECTION DRAWS

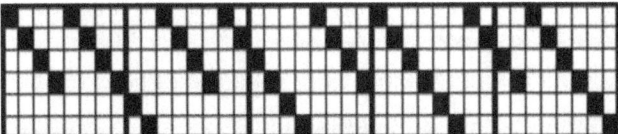

Fig. 10

[pg 12]

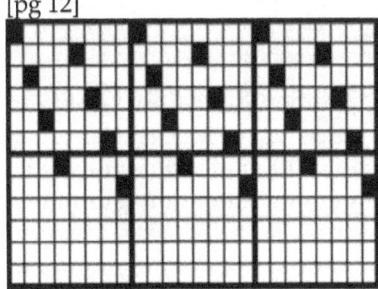

Fig. 11

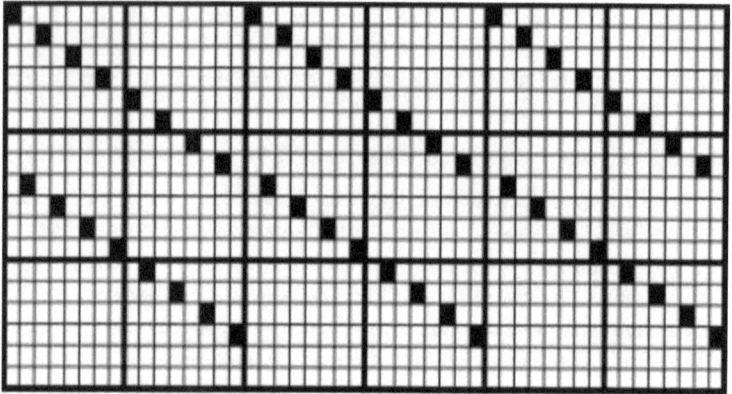
Fig. 12

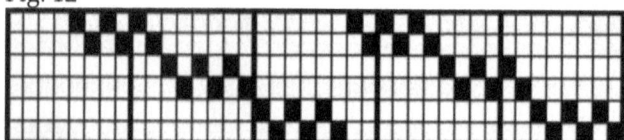
Fig. 13

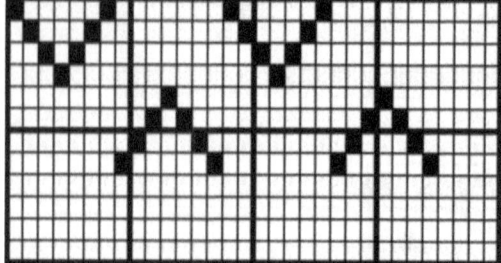
Fig. 14
[pg 13]

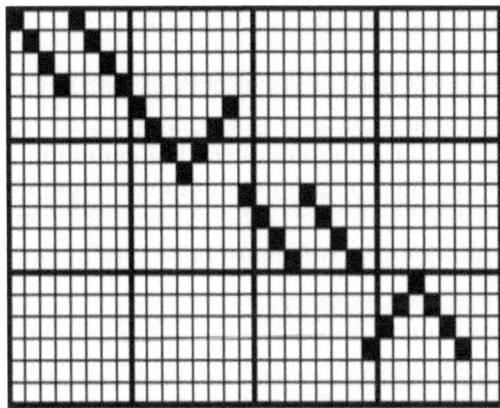

Fig. 15

This division of drawing-in drafts is used extensively in silk manufacturing; for instance, in all fabrics having a ground warp and a binder warp, also in double-face goods, or where two different weaves are combined in one effect.

One or more threads are drawn on the first section, then one or more on a second and third, if the harness is divided in so many sets.

The following examples will illustrate the principle of these draws.

In Fig. 10, shafts 1, 2, 3, 4 from the first set, shafts 5 and 6 the second, 8 threads are drawn straight on the first, then 2 on the second section.

Fig. 11, first set shafts 1 to 6 inclusive, second set shafts 7 and 8.

Fig. 12 is drawn end and end on two sections having 8 shafts each.

Figs. 13, 14 and 15, while not strictly belonging to the class of section draws, may, however, be considered under this heading. The idea is to draw a certain number of ends in one part of the harness and another group in another part, be it straight, point or skip, which will cause the effect on the cloth to be accordingly transposed or broken up.

[pg 15]

THE WEAVES AND THEIR CONSTRUCTION

In any woven fabric we distinguish two systems of threads, the *Warp or Chain*, running lengthways in the cloth, and the *Filling or Weft*, crossing the former at right angles.

This crossing or interlacing consists of every individual warp-thread being placed alternately under and over one or more threads of the filling system. The arrangement of this interlacing is technically called the *Weave*, and the variety in which the points of crossing can be distributed is practically endless.

It is principally the weave that lends to a fabric its character, influenced, of course, by the material used, the size and tension of the threads and the combination of the colors.

The weaves are divided into three main classes: *the Foundation weaves*. In the silk business they are known under the following names:

The Taffeta Weave,

The Serge Weave,

The Satin Weave.

In the foundation weaves each thread effects only one crossing in one repeat of the weave, and the points of interlacing occur in a given rotation. A repeat in the foundation weaves comprises the same number of warp-threads as of [pg 16] *picks* or filling threads, and if this number is 8, for instance, the weave is called an 8-shaft or an 8-harness weave. In marking out a weave, the warp-threads are represented by vertical lines, the filling by horizontal ones, or in each case by the space between these lines. The places where a warp-thread lies over the filling are marked with paint or simply with a cross. In a similar manner we mark out the *chain draft*, which indicates the rotation in which the shafts are raised.

[pg 17]

THE TAFFETA WEAVE

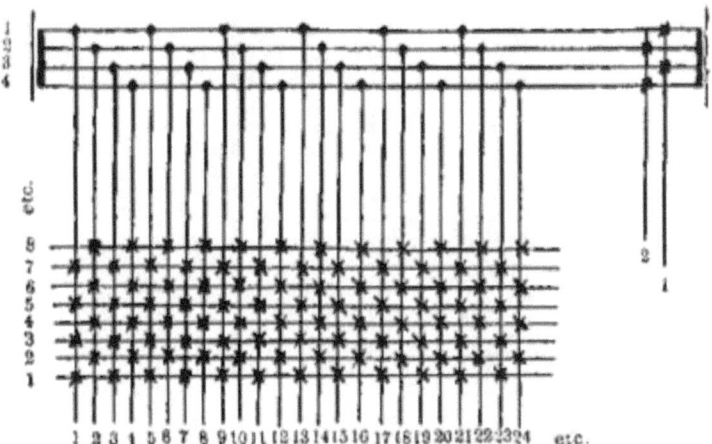

Fig. 16

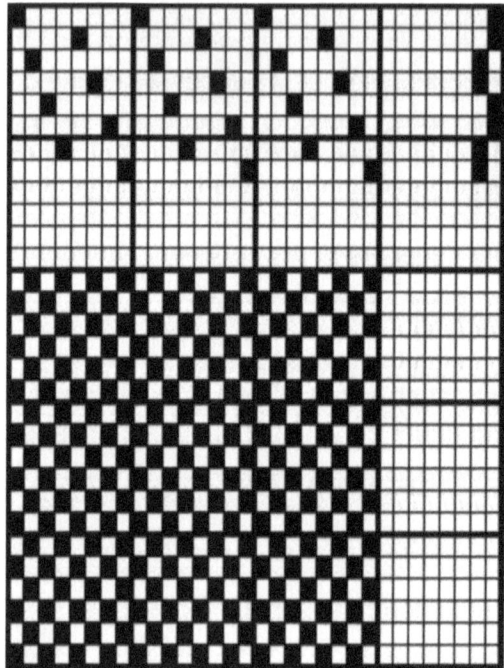

Fig. 17
[pg 18]

This is the simplest and oldest method of interlacing. The odd numbers of warp-threads cross the even numbers after every pick; hence of two warp-threads one will always go over the first pick and under the second, and the other end under the first and over the second pick. Taffeta cloth, therefore, has the same appearance on both sides, and in cotton and wool weaving this weave is technically—and properly indeed—called the *Plain Weave*. It has the smallest repeat, 2 warp-threads and 2 picks, and the exchanging of warp and filling is the most frequent possible. The cloth thus produced is firmer and stronger than that obtained with any other weave.

Fig. 16 is a taffeta on 4 shafts straight draw, the draft executed in the manner which we have already mentioned in explaining the drawing-in drafts.

Fig. 17 on common designing paper, illustrates a taffeta made on 8-harness, skip draw.

Be it mentioned that the drawing-in draft and the chain draft will be added throughout this work, the former over the weave to correspond with the respective warp-threads, the latter to the right of the drawing-in draft.

[pg 19]

GROS DE TOURS WEAVES

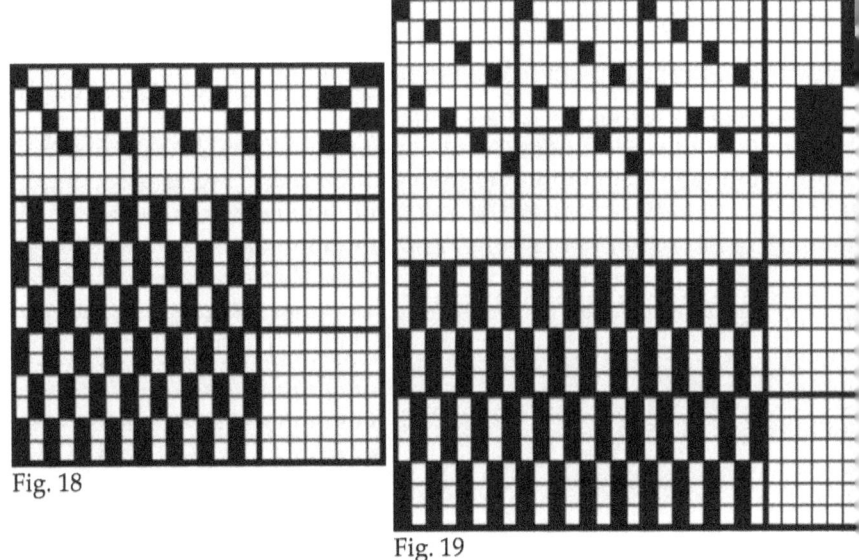

Fig. 18

Fig. 19

In this weave the working of the warp is the same as in taffeta, except that instead of one pick, two or more are inserted in the same shed. It is mostly used in selvedges, where it serves to give more firmness to the edge of an otherwise loosely woven cloth, and prevents the weaving ahead of the edge in a tight weave. Gros de Tours is sometimes used, especially when cotton or wool filling is employed, with a view to lay two picks nicely side by side, whereas a thread entered two ply with the taffeta weave will always receive some twist, which may disturb the perfect evenness of the fabric.

Fig. 18 is a Gros de Tours with two picks on four harness straight through.

Fig. 19 illustrates this weave with three picks drawn end and end on two sections of four shafts each.

[pg 20]

SERGE or TWILL WEAVES

While the taffeta weave produces either an entirely smooth fabric, or one with a distinct transverse rib as in gros-grain, the twill weave forms diagonal lines on the cloth, running either from left to right or from right to left.

To make a twill, not less than three ends and three picks are required, of which each thread floats over two of the other system and interlaces with the third. The rotation of the interlacing is always consecutive, that is it moves with each succeeding pick one thread to the right (or to the left if the lines are to run in that direction). If warp and filling have the same texture, that is the same number of threads in a given space, the twill lines will form an angle of 45°; if the warp stands closer than the filling, the incline will be steeper, and in the opposite case the angle will approach more the horizontal.

The weaves can be expressed in numbers, for instance: the 3-end twill warp effect would be marked 2-1, which indicates that each warp-thread goes over two and under one pick.

Twill weaves are called *evensided* when the arrangement of "warp up" and "filling up" are evenly balanced, and *unevensided* if either warp or filling predominate on the face of the fabric; the latter class is therefore subdivided in *Warp effects* and *Filling effects*.

In the following a number of serge weaves are illustrated, the French designations being added in some cases, as they are still extensively used in the trade.

[pg 21]

FILLING EFFECTS

Satin de Lyon, 2-1.

On 6 harness straight through.

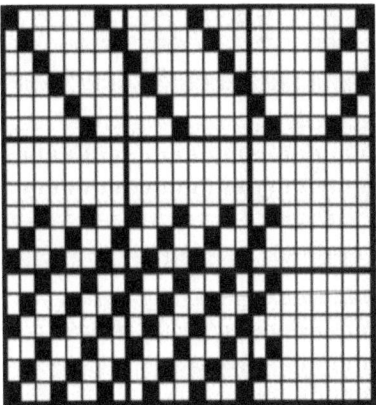

Fig. 20

Levantine, 3-1

On 8 shafts skip draw.

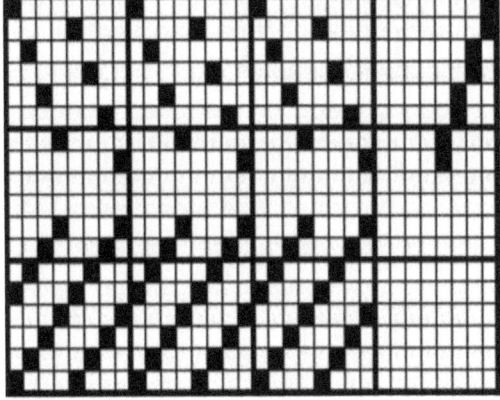

Fig. 21
[pg 22]

Polonaise, 5-1

On 12 harness skip draw.

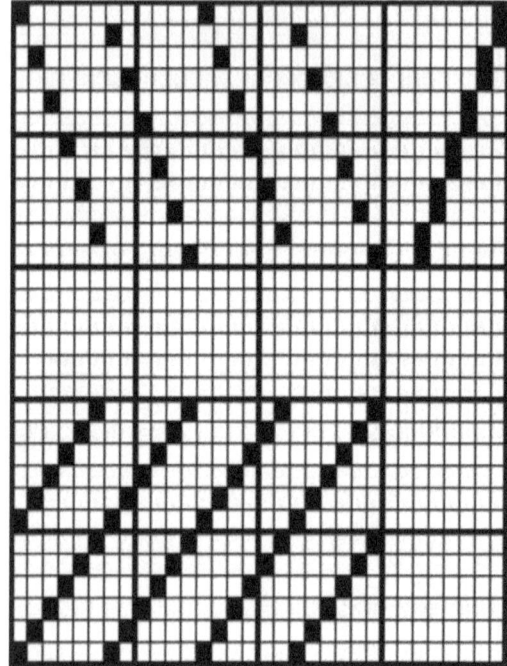

Fig. 22

Serge grosse côte, 7-1

On 8 shafts straight through.

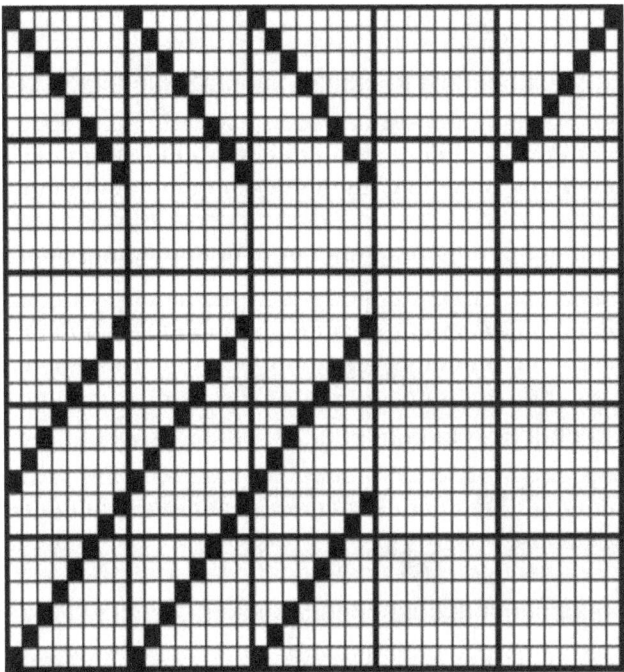

Fig. 23
[pg 23]

Serge remaine, 6-2

On 8 shafts skip draw.

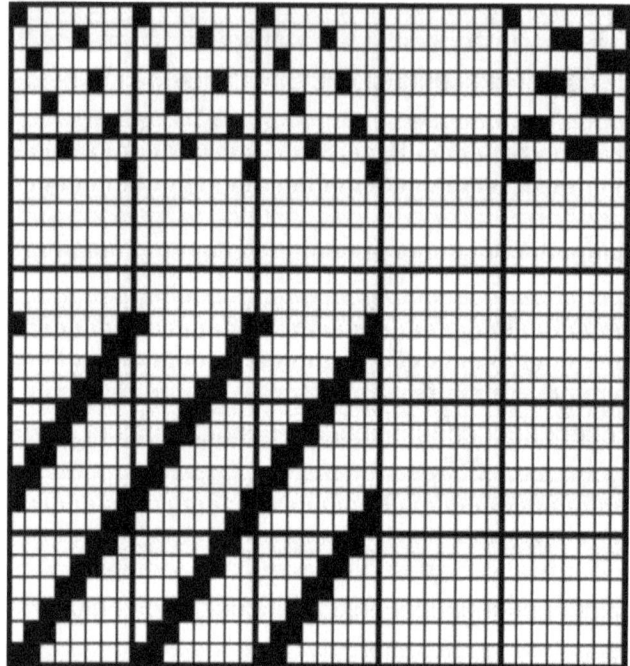

Fig. 24

Serge, 5-1, 1-1.

On 8 shafts skip draw.

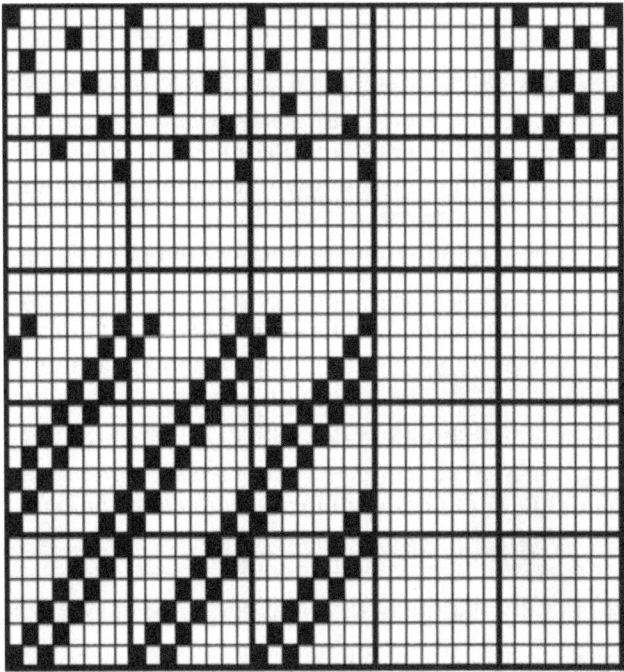

Fig. 25
[pg 24]

Serge, 4-2, 1-1, 1-1.

On 10 harness straight through.

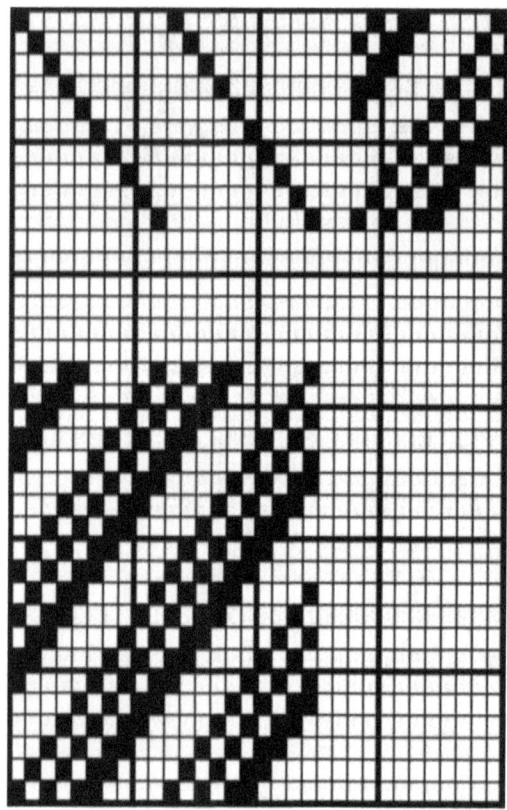

Fig. 26

Serge, 7-1, 1-1, 1-1, 1-1, 1-1.

On 16 shafts skip draw.

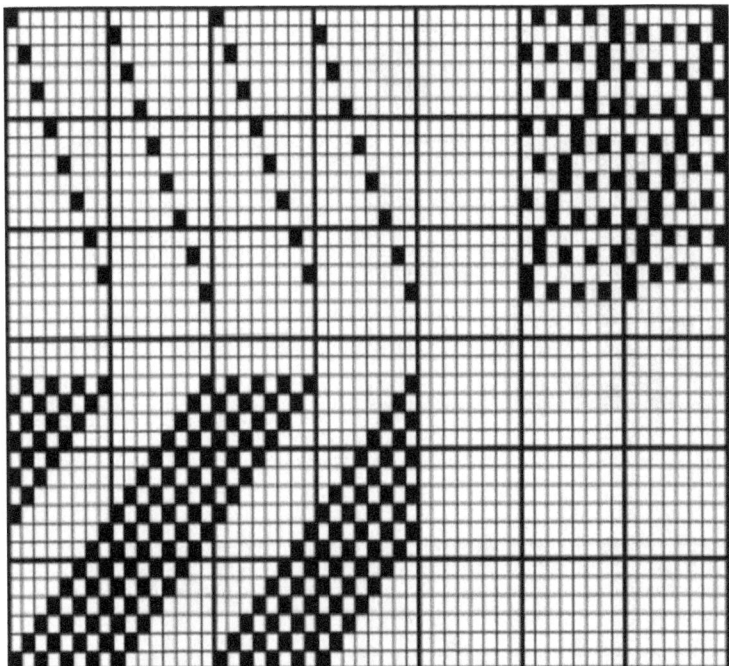
Fig. 27
[pg 25]

WARP EFFECTS

Levantine, 3-1.

On 4 shafts straight through.

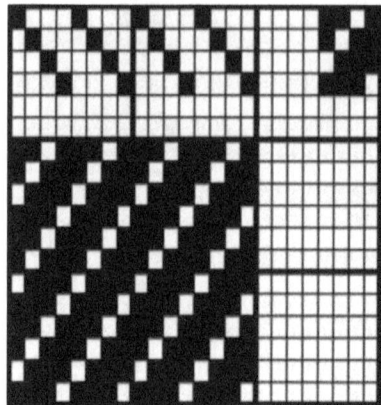

Fig. 28

Serge, 5-1, 1-1.

On 8 shafts skip draw.

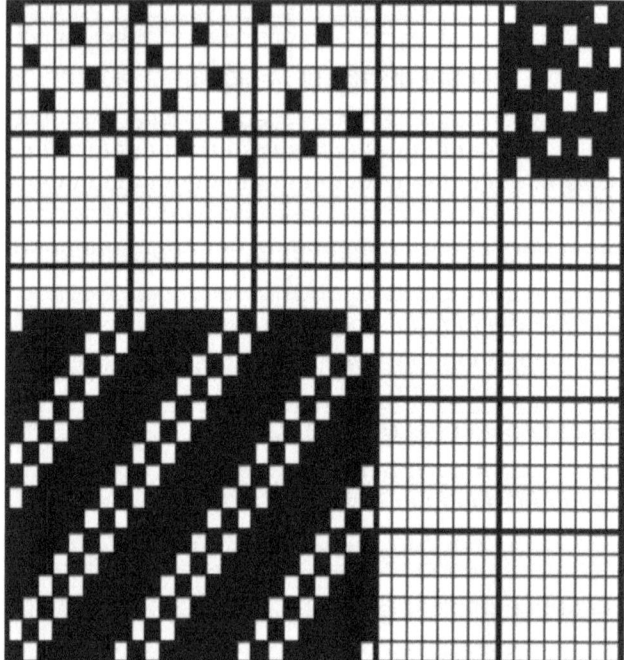

Fig. 29

EVENSIDED TWILLS

Surah, 2-2

On 4 shafts straight through.

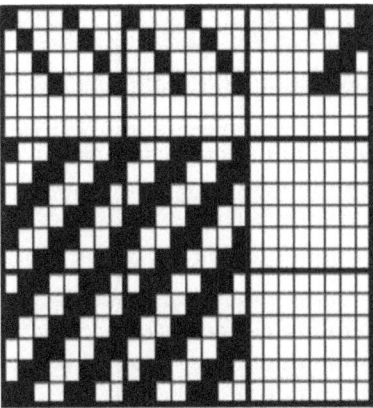

Fig. 30

Croise, 3-3, 2-2, 1-1.

On 12 shafts straight through.

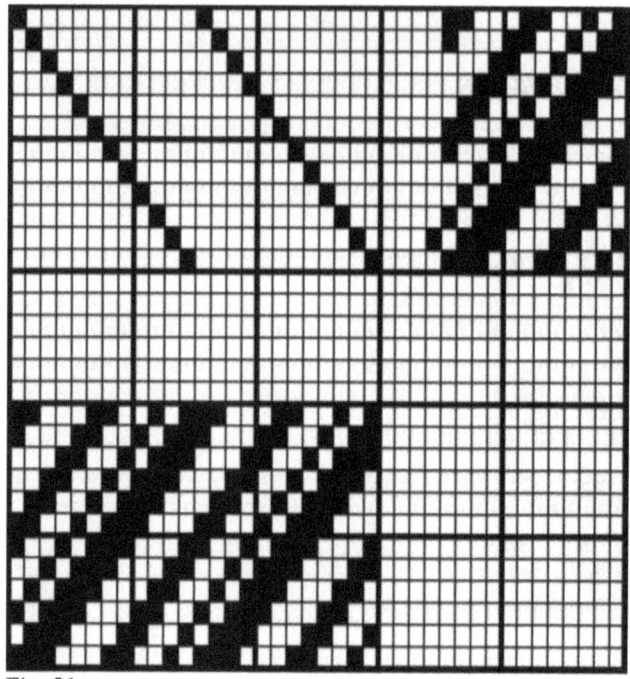

Fig. 31
[pg 27]

Serge, 7-7, 1-1, 1-1, 1-1

On 20 shafts skip draw.

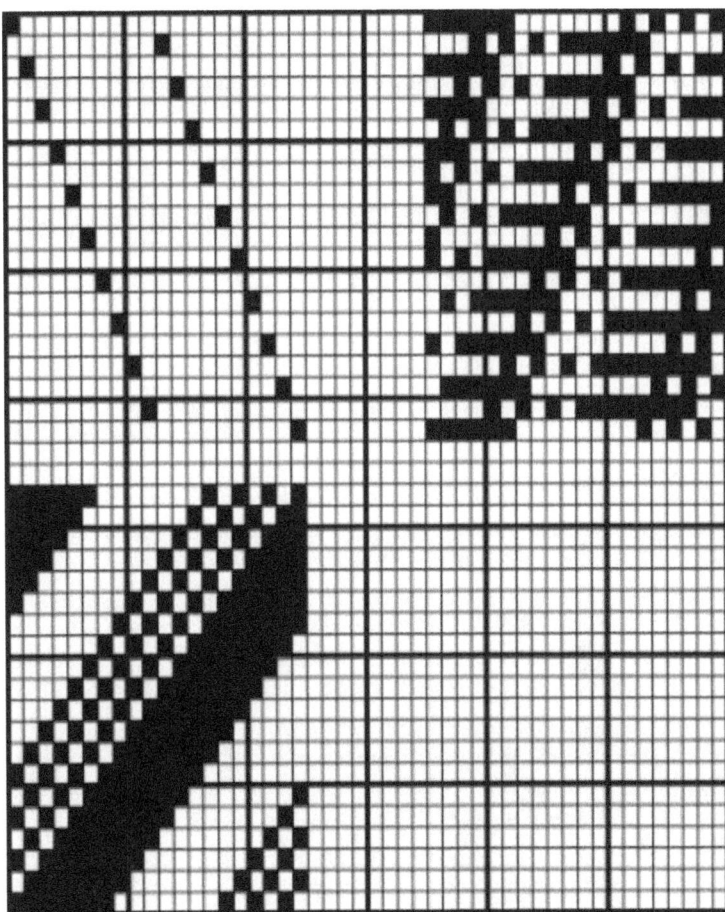

Fig. 32

POINTED TWILLS

in the direction of the filling and also of the warp.

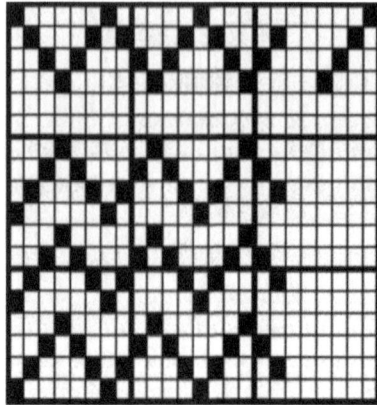

Fig. 33

On 4 shafts point draw, weave 3-1.

[pg 28]

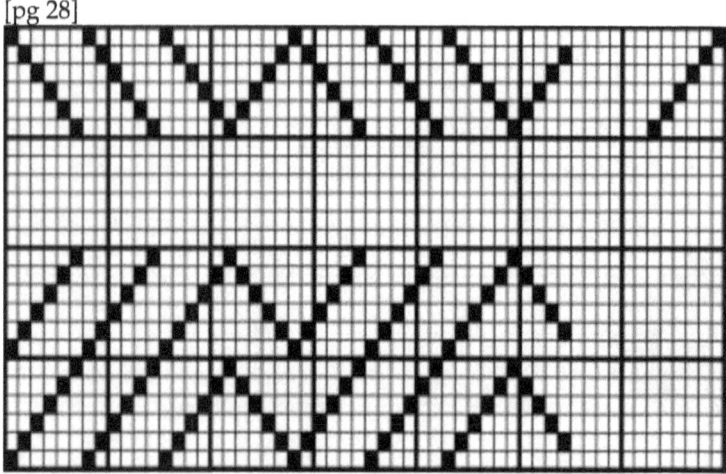

Fig. 34

On 6 shafts point draw, drawn as follows:

Eighteen ends from back to front and 4 ends from front to rear, weave 5-1.

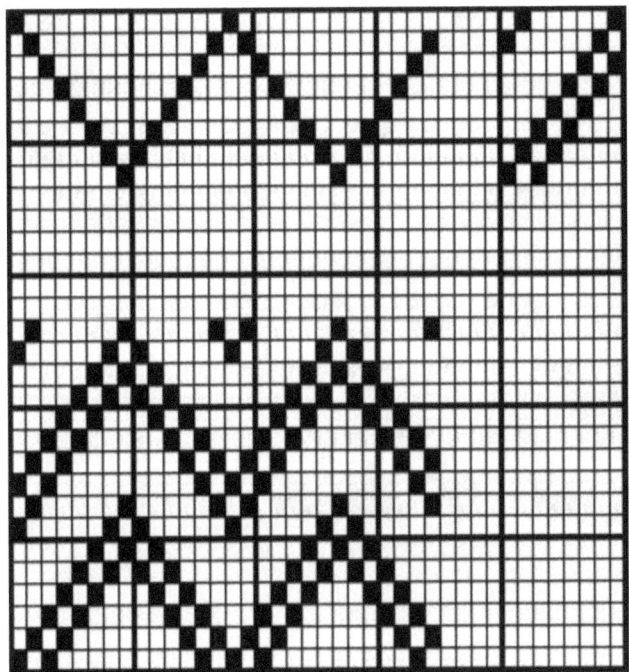

Fig. 35

On 8 shafts pointed draw, weave 5-1, 1-1.

[pg 29]

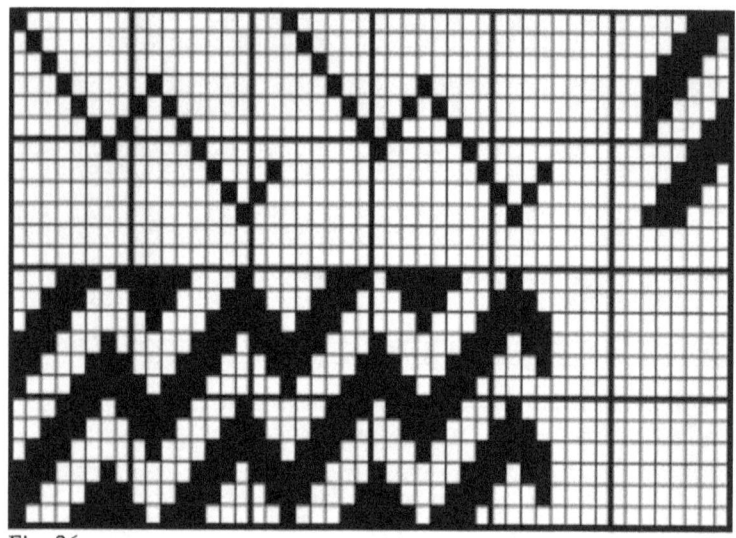

Fig. 36

On 10 shafts, with weave 3-3, drawn as follows:

7	threads	from	back to front.
3	"	"	front to rear.
6	"	"	back to front.
2	"	"	front to rear.

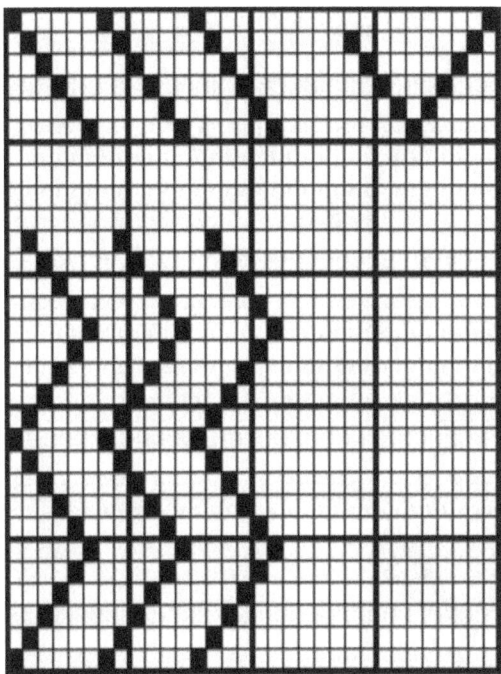

Fig. 37

On 6 shafts straight draw, pointed weave 5-1.

[pg 30]

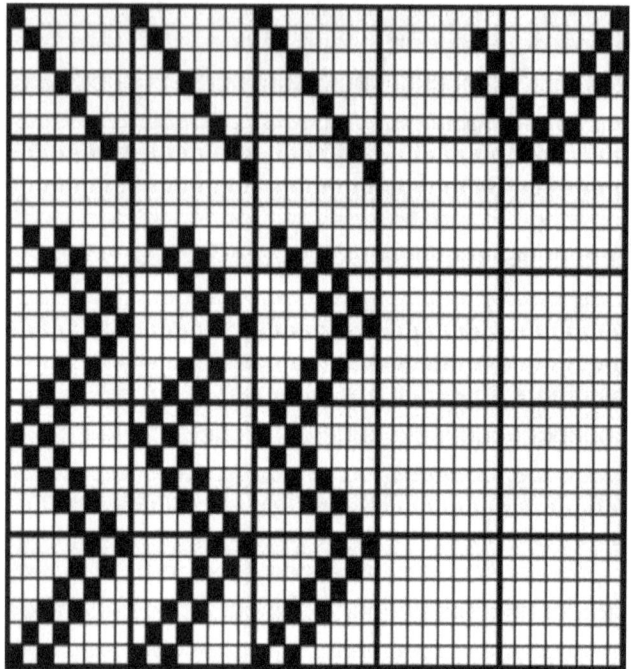

Fig. 38

On 8 shafts straight through, pointed weave 5-1, 1-1.

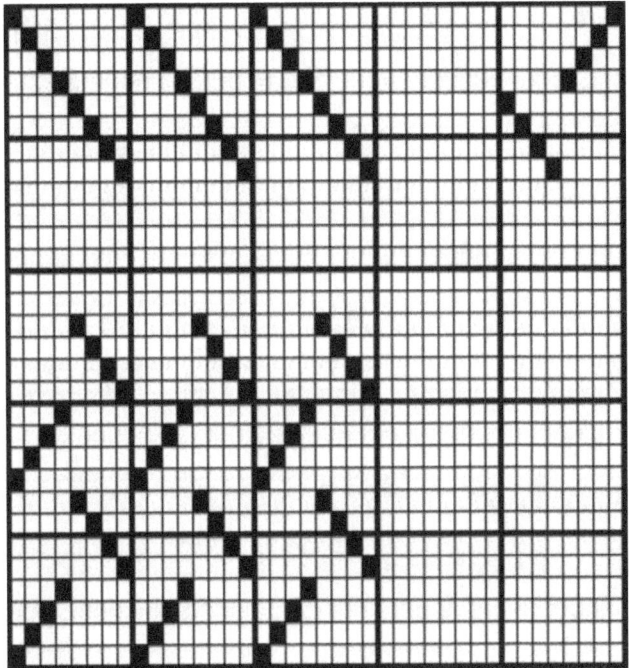

Fig. 39

Broken pointed twill, on 8 harness.

[pg 31]

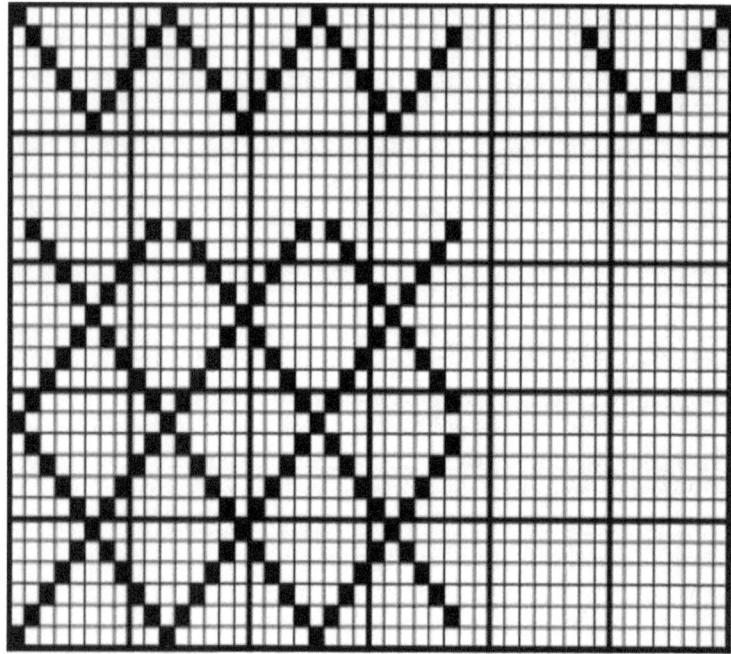

Fig. 40

On 6 shafts point draw, pointed weave 5-1.

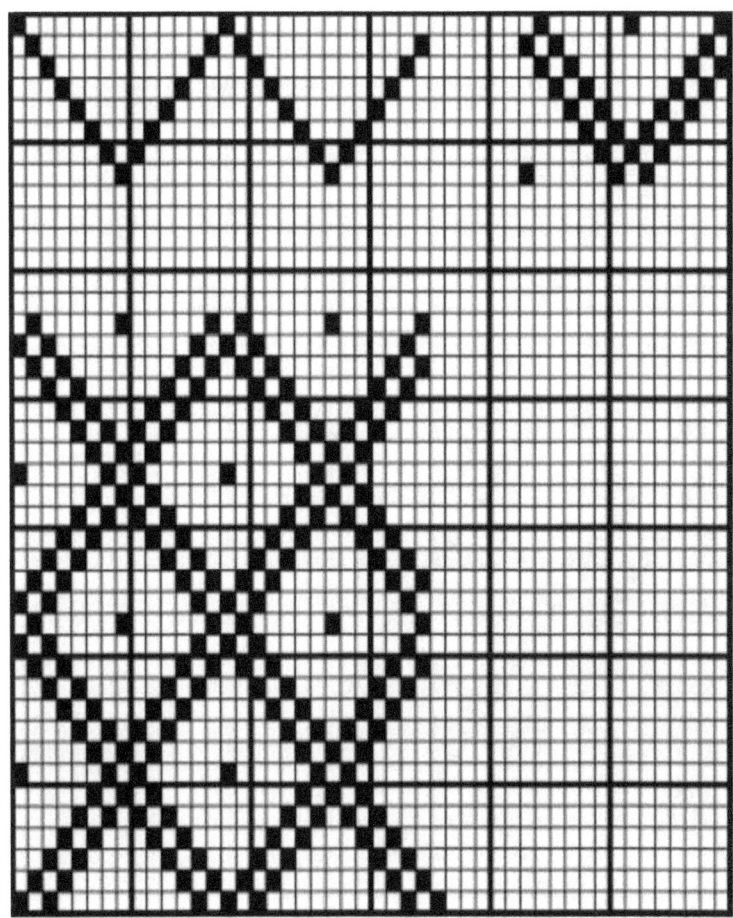

Fig. 41

On 8 shafts point draw, pointed weave 5-1, 1-1.

[pg 32]

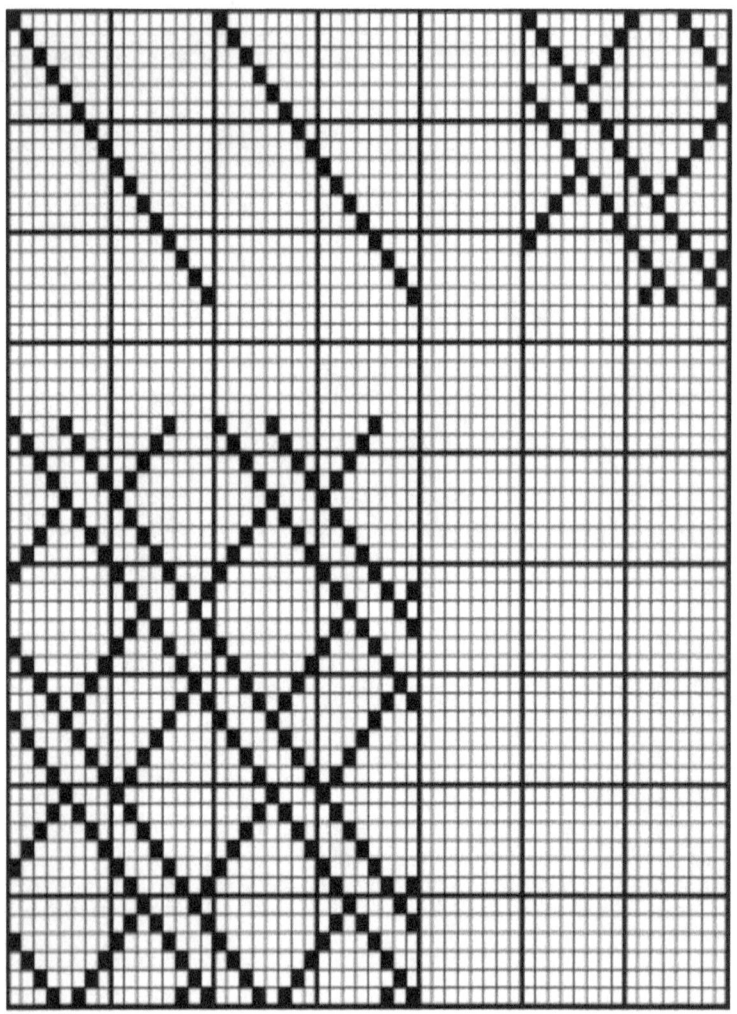

Fig. 42

Fancy twill, on 16 shafts straight draw.

SATIN WEAVES

The distinct diagonal lines which characterize the class of weaves explained in the previous chapter are absent in the satin weaves;

and while the interlacing in the former is done in a strictly consecutive order, we endeavor to scatter the points of stitching in the latter as much as possible, in order [pg 33] to create an entirely smooth and brilliant surface on the cloth.

In all satins the number of ends in a repeat is the same in warp and filling.

The lowest repeat of a regular satin comprises five threads of each system, and the interlacing is done in the following order:

The	1st	pick	with the	1st	warp-thread
"	2d	"	"	3d	"
"	3d	"	"	5th	"
"	4th	"	"	2d	"
"	5th	"	"	4th	"

Fig. 43 illustrates this weave. An examination of the rotation, as given above, will show that every warp-thread intersects two picks apart from its neighbor. The number "2" is in this case what is technically known as the *counter*, that is the number which indicates the points of interlacing by adding it to number 1 and continuing so until all the warp-threads are taken up.

The following is the rule to find the counter for any regular satin:

Divide the number of harness into two parts, which must neither be equal nor have a common divisor. Any of these two numbers can be used for counting off, but usually the smaller one is taken. According to this rule we obtain a regular satin

On	5	harness	with counter	2
"	7	"	"	2 or 3
"	8	"	"	3
"	9	"	"	2 or 4
"	10	"	"	3
"	11	"	"	2, 3, 4 or 5
"	12	"	"	5
"	13	"	"	2, 3, 4, 5 or 6

"	14	"	"	3 or 5
"	15	"	"	2, 4 or 7
"	16	"	"	3, 5 or 7.

[pg 34]

The 4 harness broken twill, Fig. 53, is sometimes classed among the satins.

The 6 harness satin, Fig. 54, is irregular; as a counter cannot be derived from number 6 by the given rule. The rotation generally used is 1, 3, 6, 4, 2, 5.

Regular Satins

5 harness Satin, "Satin de Chine."

Straight draw, counter 2.

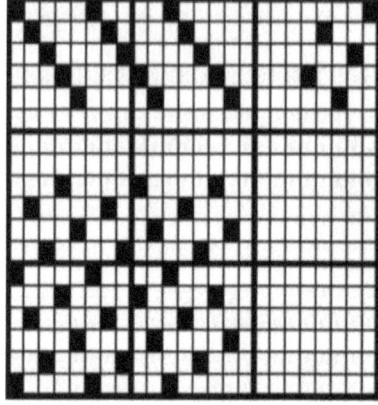

Fig. 43

7 harness Satin, "Satin Merveilleux."

Skip draw, counter 2.

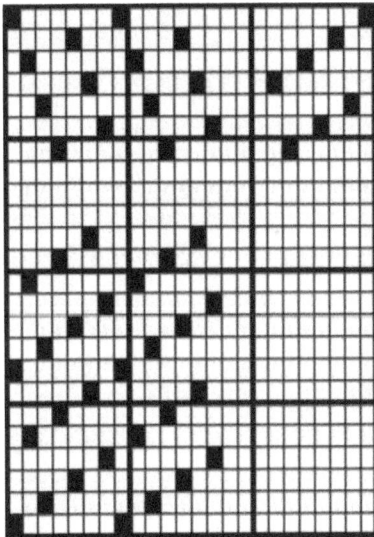

Fig. 44
[pg 35]

7 harness satin

Straight draw, counter 3.

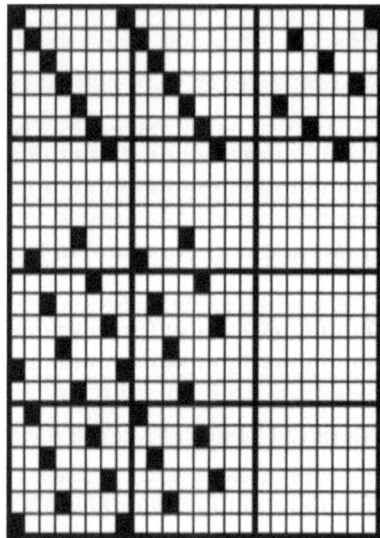

Fig. 45

8 harness satin "Duchese"

Skip draw, counter 3.

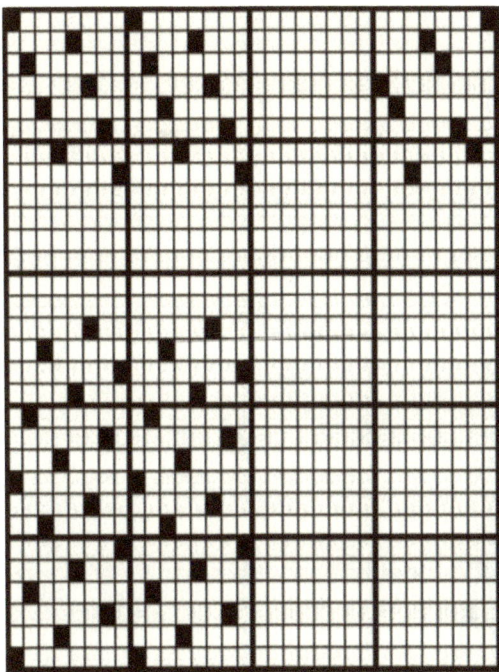

Fig. 46
[pg 36]

9 harness satin

Straight draw, counter 4.

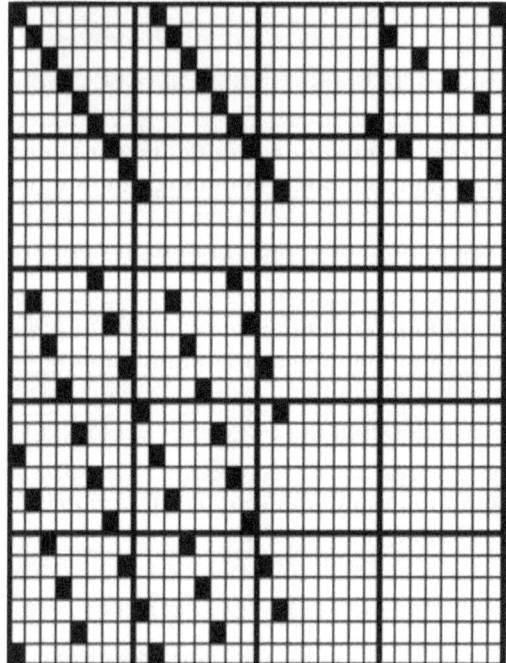

Fig. 47

10 harness satin

Straight draw, counter 3.

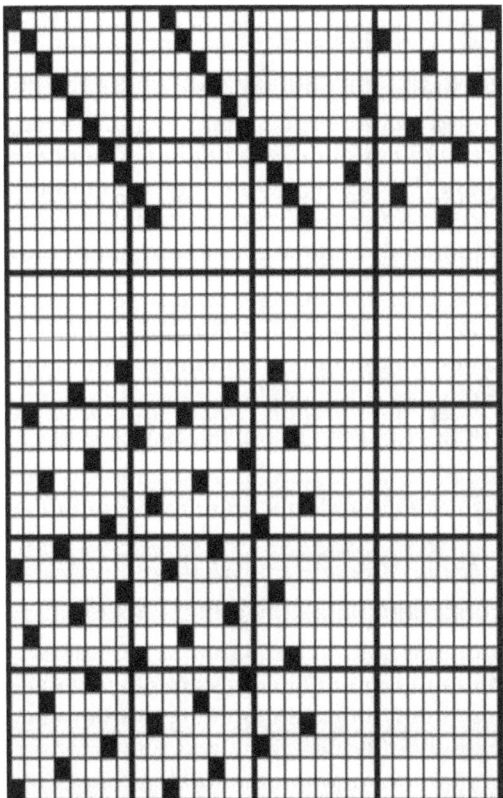

Fig. 48
[pg 37]

11 harness satin

Skip draw, counter 5.

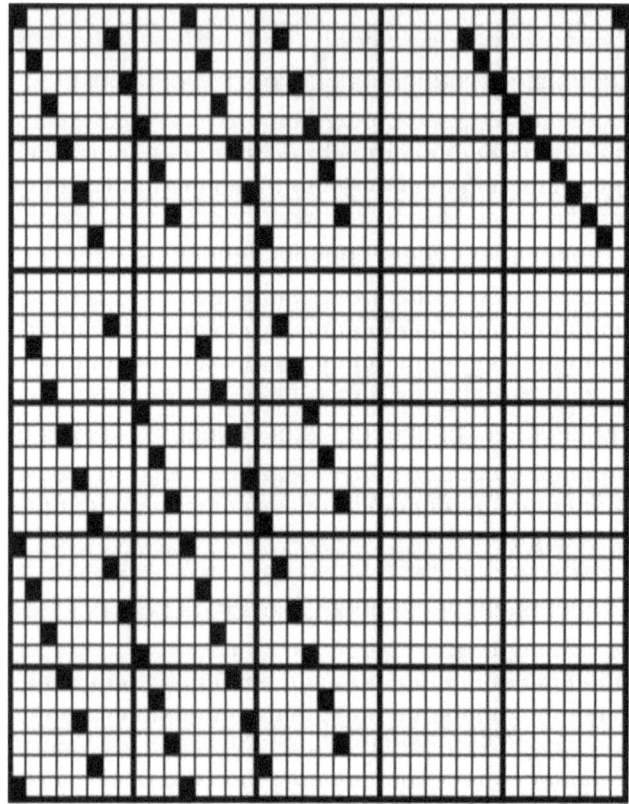

Fig. 49

12 harness satin

Skip draw, counter 5.

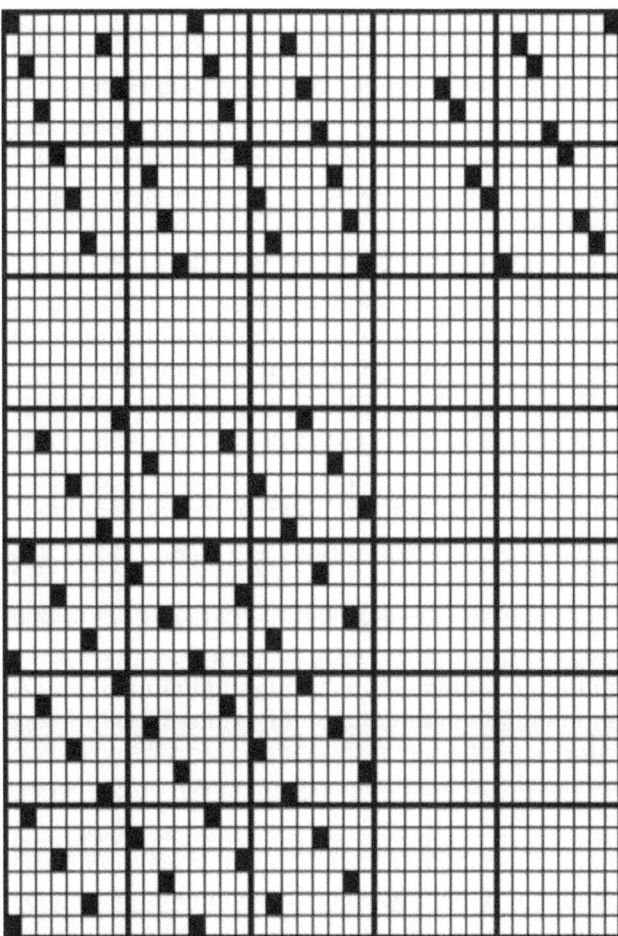

Fig. 50
[pg 38]

16 harness satin

On 2 sections of 8 shafts each, drawn end and end, counter 7.

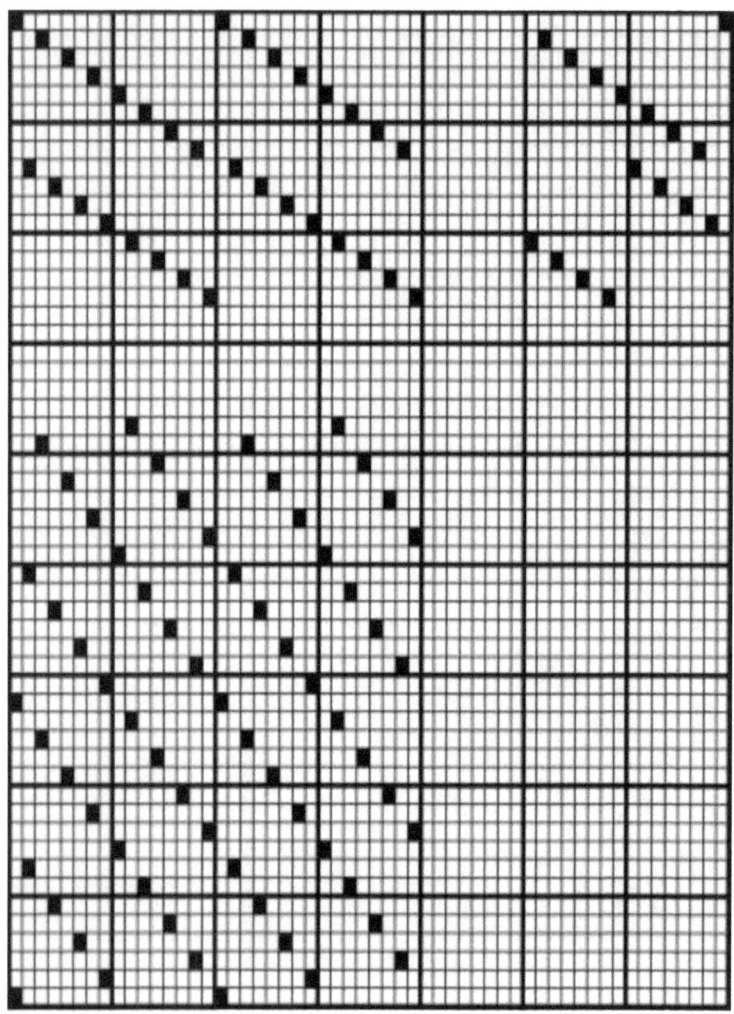

Fig. 51
[pg 39]

8 harness satin, warp effect.

Straight draw, counter 3.

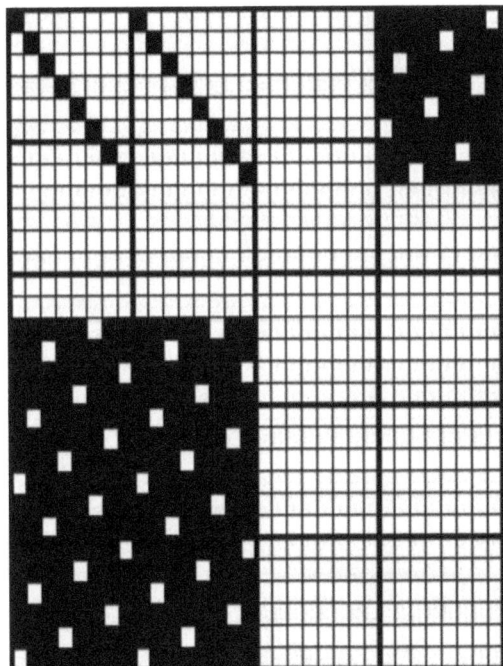

Fig. 52

IRREGULAR SATINS

Satin Turc.

On 4 shafts straight through.

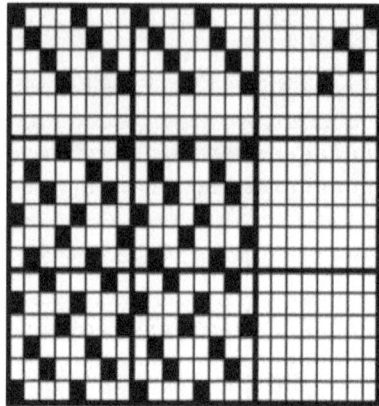
Fig. 53
[pg 40]

Satin à la Reine

On 6 shafts straight draw.

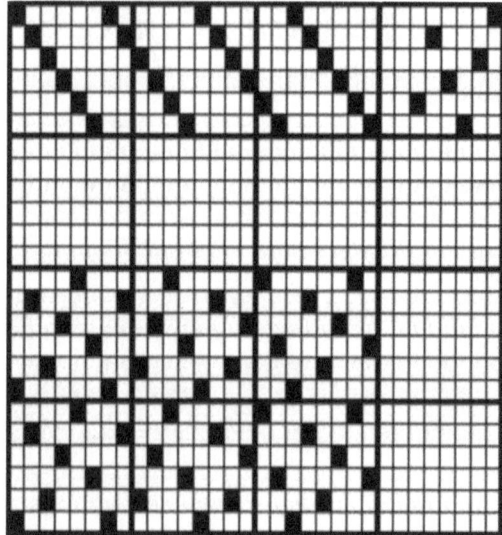
Fig. 54
[pg 41]

DERIVATIVE WEAVES

1. From the Taffeta

Royale is a modification of the regular Gros de Tours, inasmuch as the rib line, which in the latter runs straight across the cloth, is broken off after a given number of warp-threads. These groups, which may comprise 8, 12 or more threads, will interlace each one pick higher than the preceding one.

Royale of 8 ends

On 2 sections of 4 shafts each.

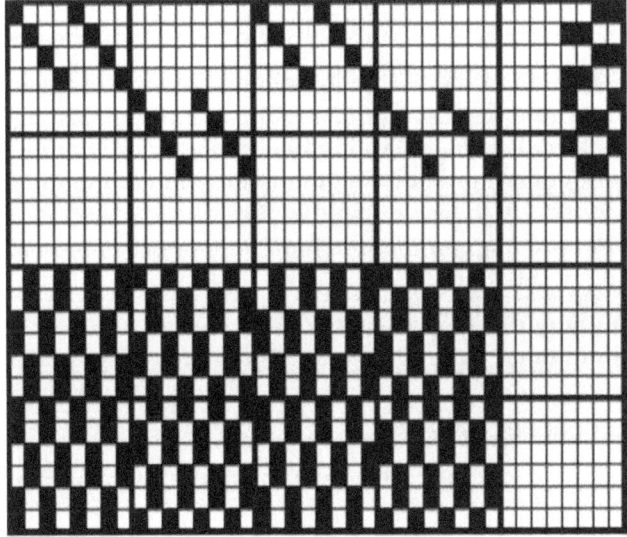

Fig. 55

Velours Ottoman or Faille française.

In order to obtain a broader rib than that of Gros de Tours, and at the same time to lend firmness to the fabric, we add to the ground warp, which forms the ribs, another [pg 42] or binder warp, which works continually taffeta, while the ground warp changes only every 3 or 4 picks for the rib.

Faille française.

4 ends of ground on the first section of 8 shafts, skip draw.
1 " binder " second " 2 "

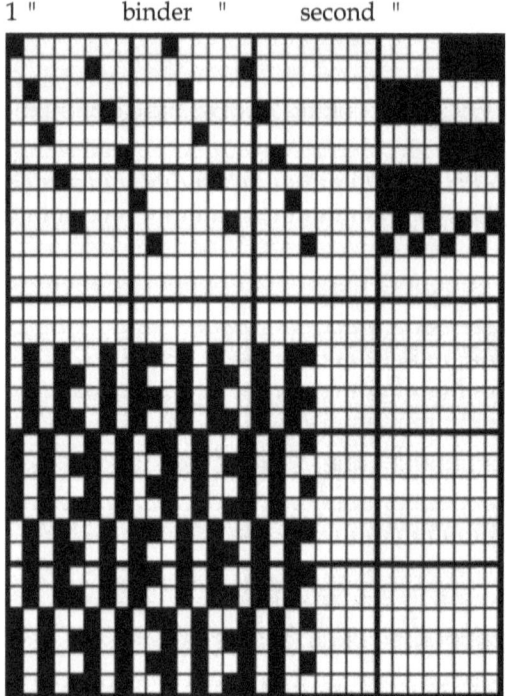

Fig. 56

Velours Ottoman without a Binder-warp.

In this weave, of which Fig. 57 illustrates a specimen, comprising 8 warp-threads and 32 picks in a repeat, the rib contains 4 picks. Of the 8 warp-threads, 3 float over and 3 under the rib, while the 2 others bind taffeta, which latter function is executed by 2 other threads in the next rib.

[pg 43]

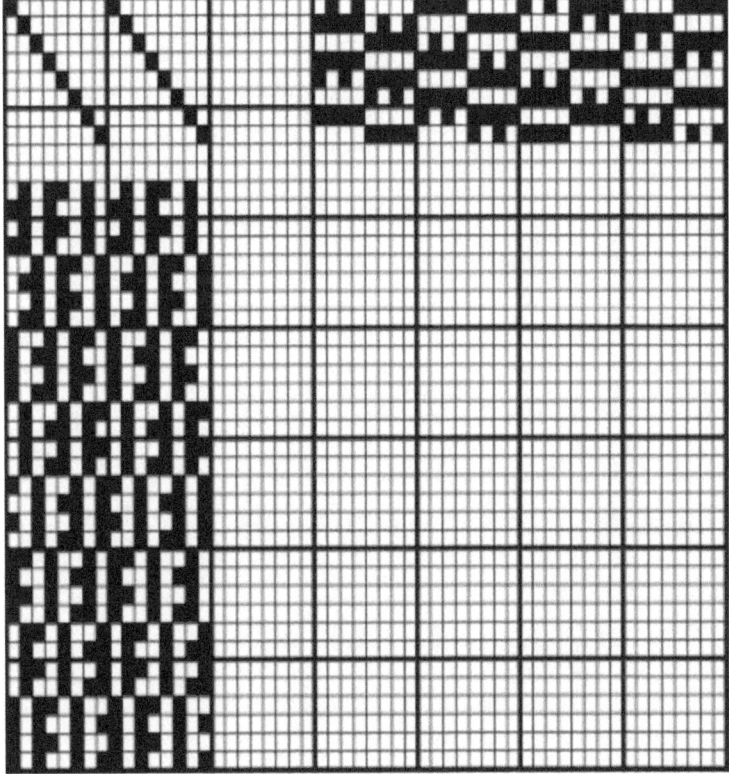

Fig. 57

2. From the Twills

One variety is obtained by interlacing the warp-threads alternately one or more picks behind, and then a number of picks ahead of their respective neighbors; so the complete arrangement of the points of binding in a repeat will generally form two parallel diagonal lines. This will cause the twill lines to appear less pronounced than is the case in the regular twill, and the character of the fabric approaches more that of the satin.

[pg 44]

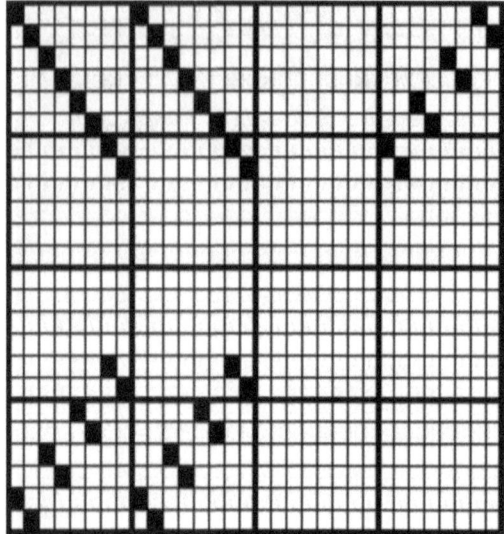

Fig. 58

Satin Sergé.

On 8 shafts, straight draw.

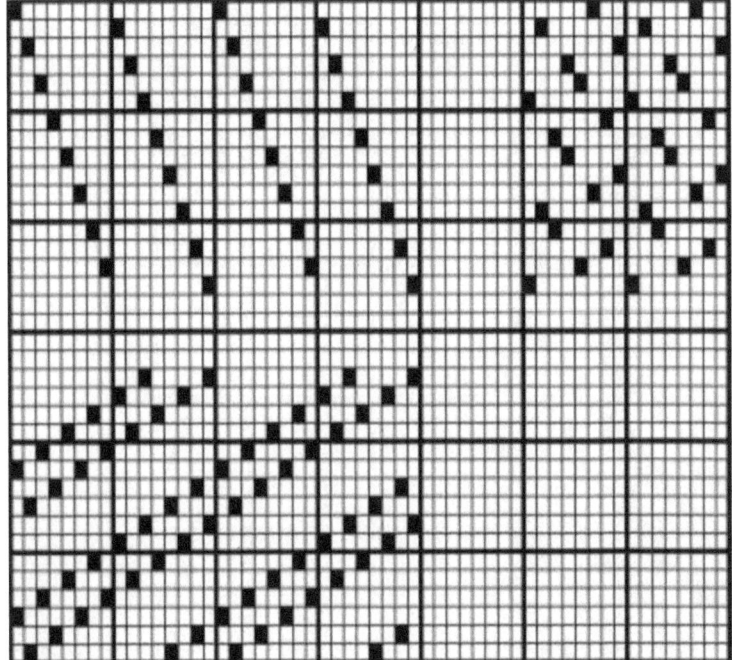

Fig. 59

Côte Satinée.

On 16 harness, skip draw.

Rhadzimir-Surah 2-2. After a certain number of picks of the regular surah all the warp-threads are crossed in two's thereby causing a sort of a rib or cut line across the fabric.

[pg 45]

Rhadzimir of 4 picks.

On 8 shafts, straight draw.

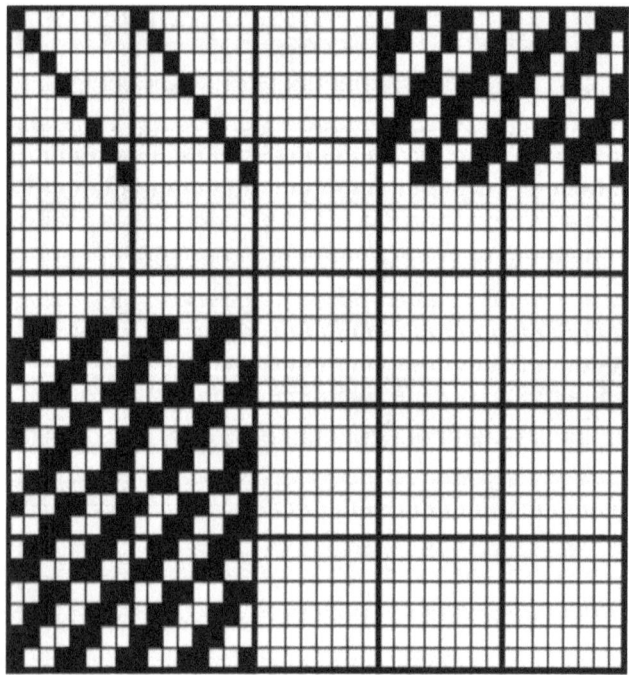

Fig. 60

Rhadzimir of 6 picks.

On 8 shafts, straight draw.

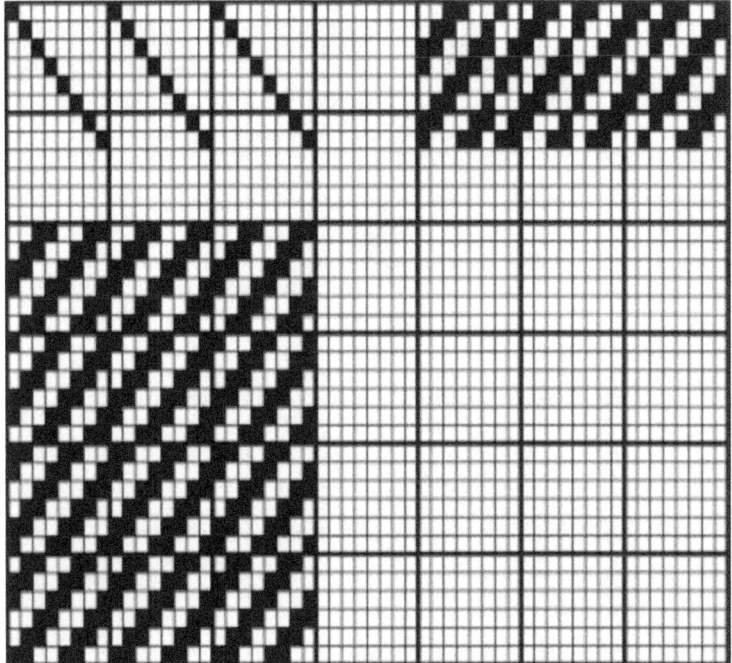

Fig. 61
[pg 46]

3. From the Satin Weave

Satin Soleil shows a satin-like surface with a cross line appearance. Fig. 62 illustrates it as made on 8 shafts, straight draw.

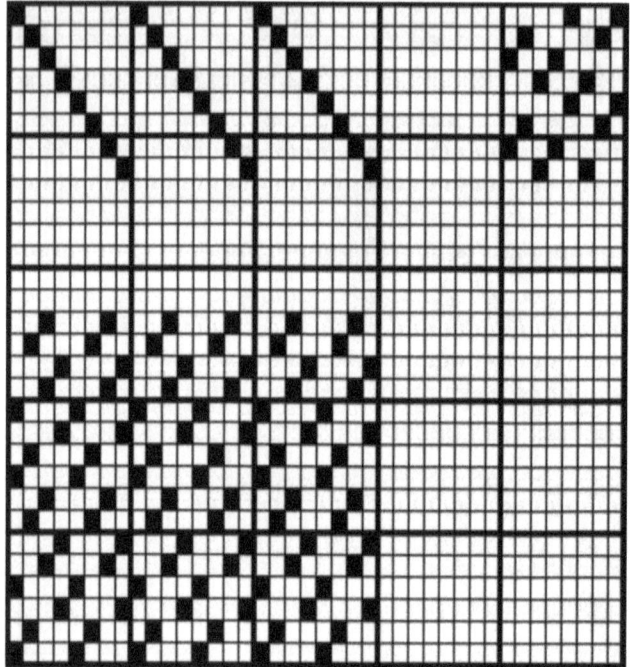

Fig. 62

Satin Grec is a 12-harness satin, in which a taffeta point is added to each place of interlacing, thus giving the cloth a much firmer hand. Fig. 63 represents this weave on 12 shafts, skip draw.

[pg 47]

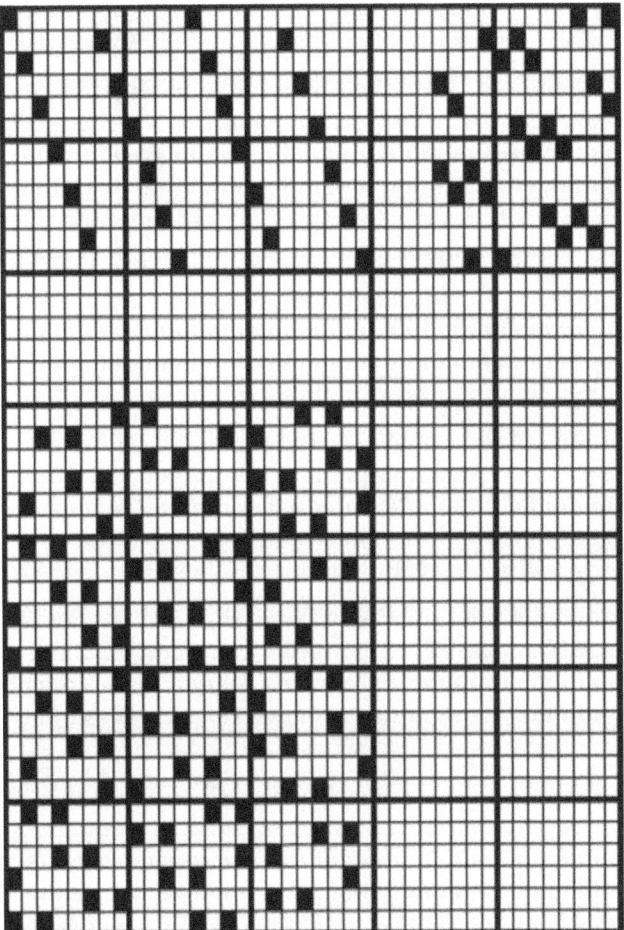

Fig. 63

Peau de Soie. An 8-shaft satin with one point added on the right or left to the original spots, giving the fabric a somewhat grainy appearence. Fig. 64 represents a peau de soie on 8 shafts, straight through.

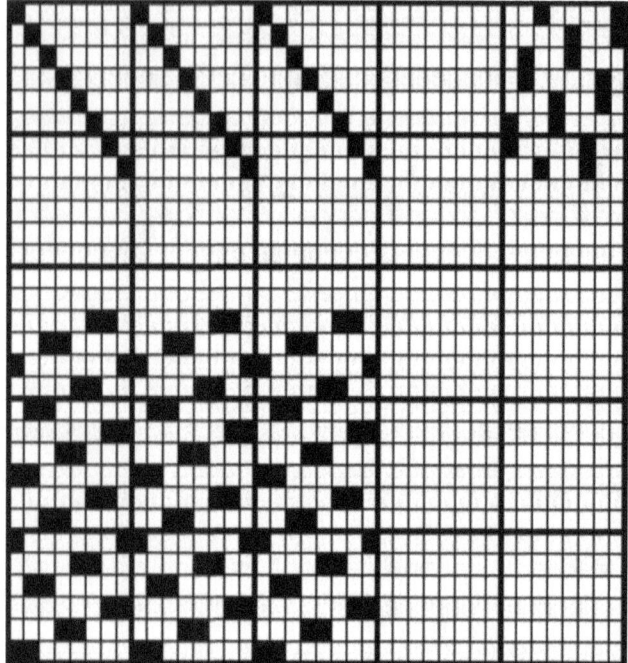

Fig. 64
[pg 48]

Fleur de Soie. The face is a satin de Lyon (2-1 twill), with a backing interlaced on the 12-shaft satin principle, Fig. 65, on 12 shafts, skip draw.

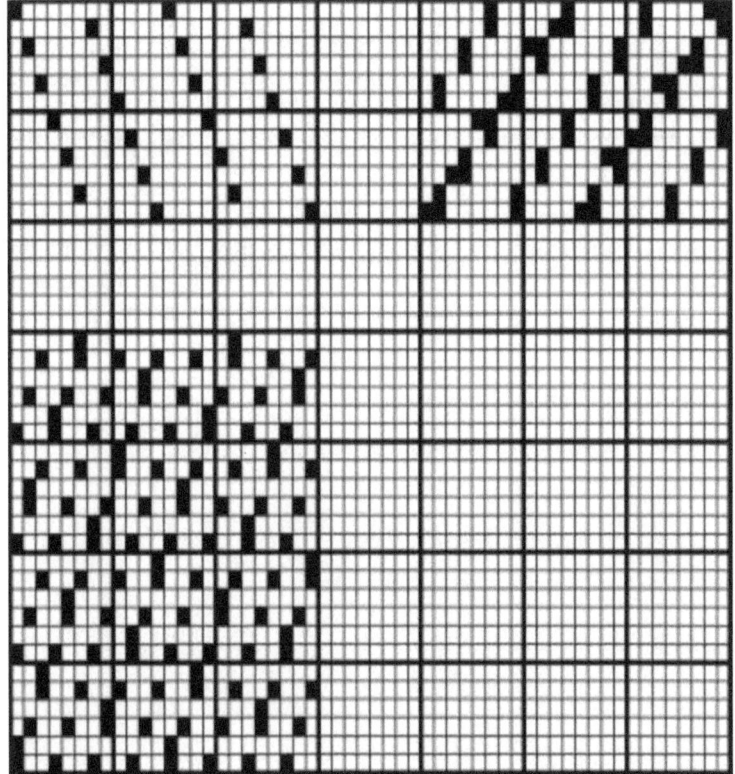

Fig. 65
[pg 49]

CANNELE and REPP WEAVES

are in their construction related to the taffeta, and are used mostly in the form of stripes as an additional ornament to a fabric. The threads going into the composition of these effects exchange continually from taffeta interlacing to floating over a certain number of threads, and must be introduced either in warp or filling close enough to make the floats cover up the taffeta work entirely, and thus enable the material used to show up with the full brilliancy it possesses.

Cannele effects can be produced in two distinct ways. One is to let every individual thread work alternately taffeta and float, while in the other method one thread weaves always taffeta, and a second thread is used for the cannele exclusively. These latter threads must come from a separate warp, which is introduced to embellish the ground or taffeta part of the fabric.

The floating threads can either stitch all on one pick and so form a continuous cut line, or be divided in groups, of which one will bind in the middle of the floats of the other group. The following designs show both the face and backside of the respective weaves:

[pg 50]

Alternating Cannele of 6 picks.

On 4 shafts, straight through.

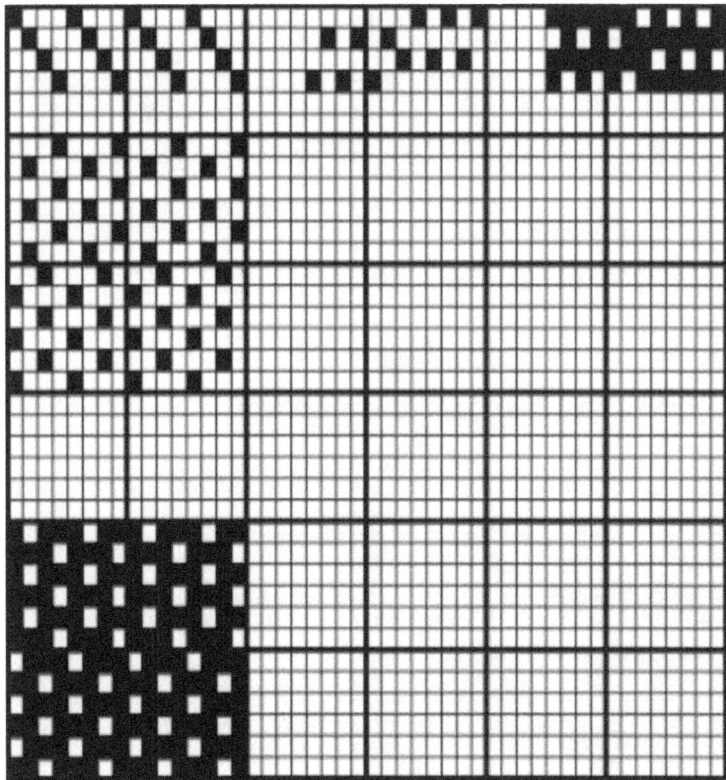

Fig. 66

Canelle (2 beams). Over 3 picks, interlacing on every fourth pick, drawn end and end on 2 sections of 4 shafts each.

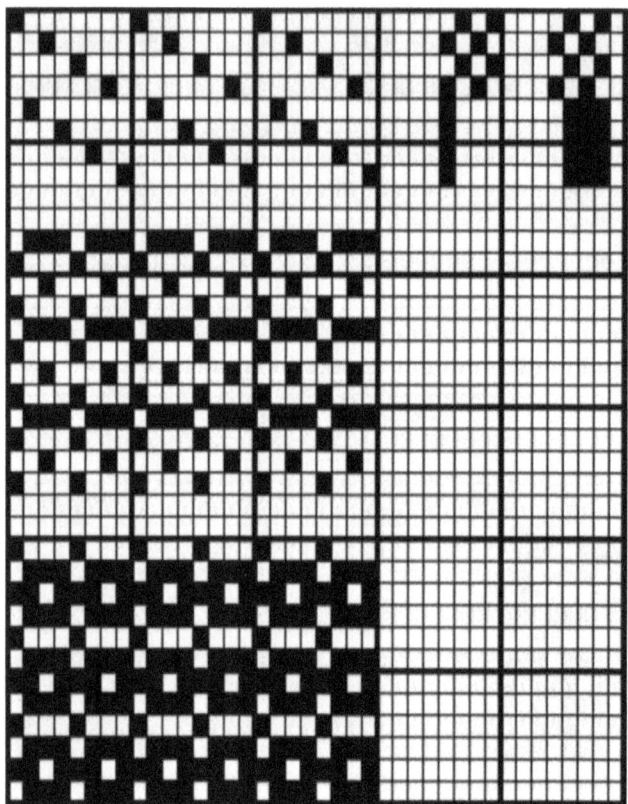

Fig. 67
[pg 51]

Cannele over 5 picks, binding on the sixth, but every second thread advanced 3 picks (to the middle of the float of the first thread), drawn end and end on 2 sections of 4 shafts each.

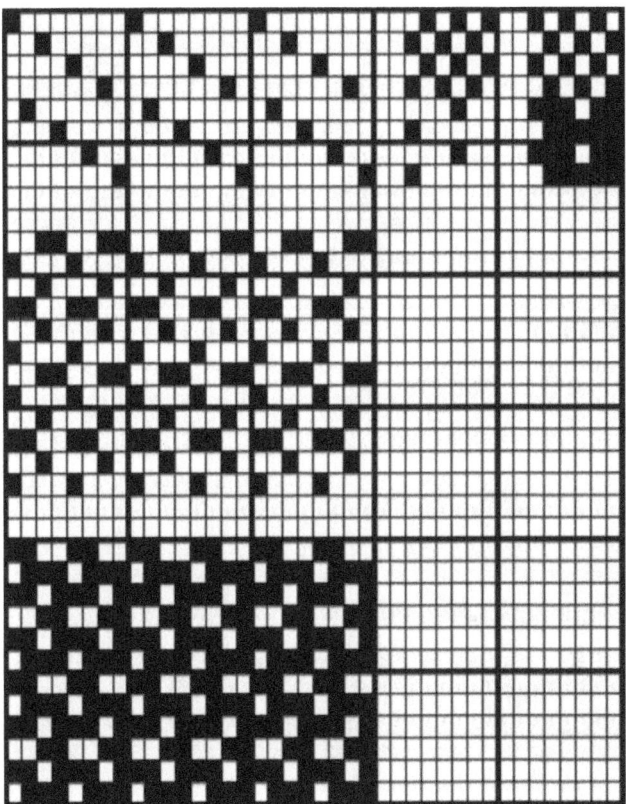

Fig. 68
[pg 52]

Cannele arranged in groups of 8 threads, floating over 6 picks and binding on the seventh and eighth, drawn on 2 sections, with 4 shafts in first and 2 in second section.

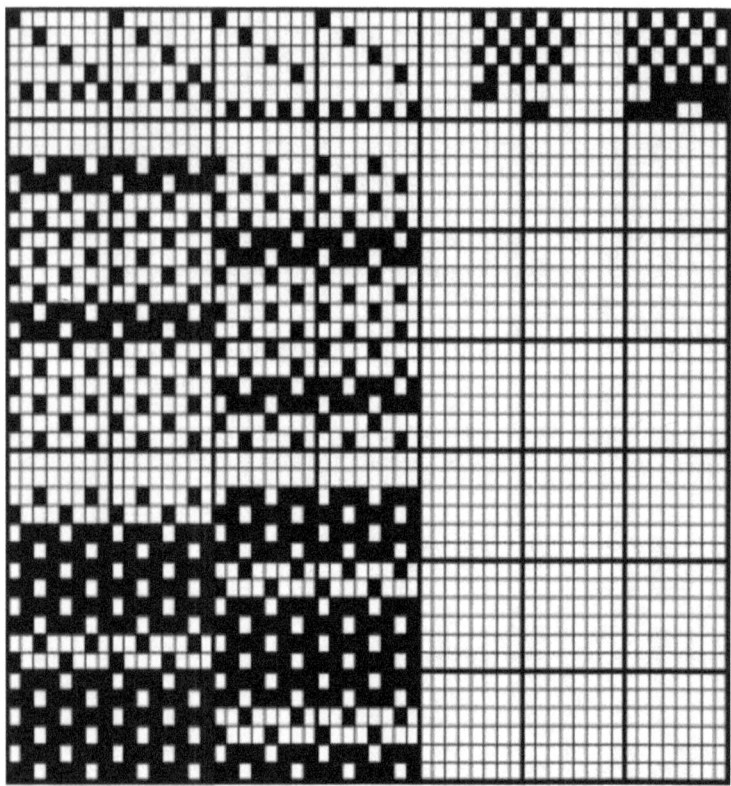

Fig. 69

Repp on 8 shafts straight through. Rotation of filling. 1 pick taffeta, 1 pick float (rib).

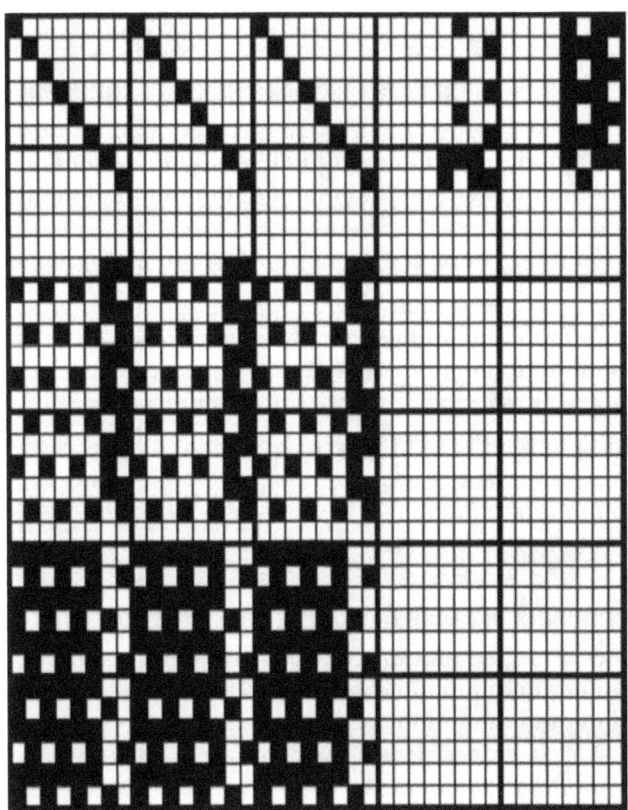

Fig. 70
[pg 53]

Repp of 8 threads, on 2 sections of 4 shafts each, 8 ends per section.

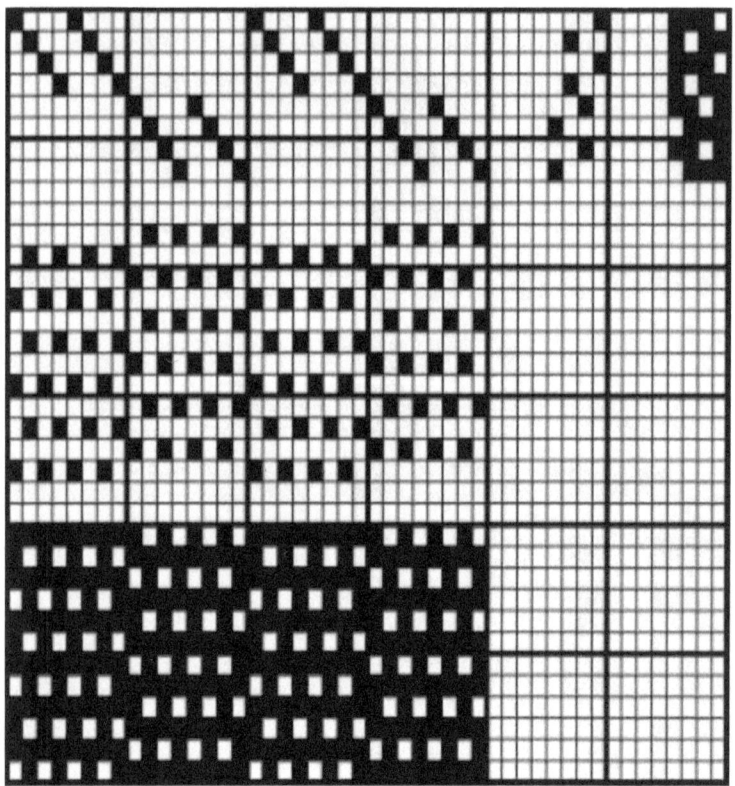

Fig. 71

Repp of 5 threads, binding on the sixth; every second pick binds on the middle of the first pick. On 6 harness straight draw.

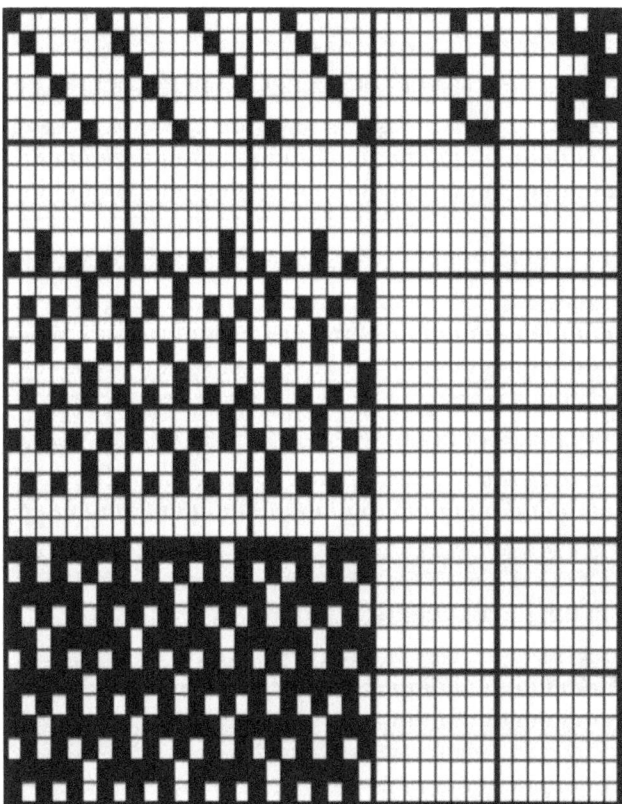

Fig. 72
[pg 54]

Repp in groups, floating over 6 ends and binding on the seventh and eighth on 8 shafts straight draw.

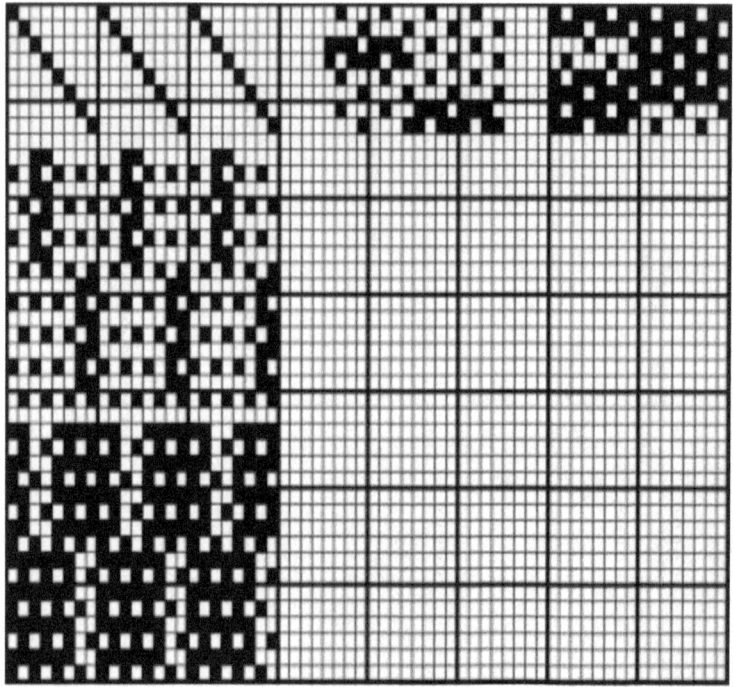

Fig. 73
[pg 55]

DOUBLE FACED FABRICS

In this class we find either two systems of warp or of filling so combined that only one will be visible on either side. The color on one side is generally different from the other, and so may the interlacing be of a different nature on face and back. In the latter case great care must be exercised not to allow the weave on one side to disturb the one on the other, and as a rule the points of interlacing of the first warp or filling system are placed as much as possible in the middle of the floats of the second. This will prevent either color or weave to be seen on the opposite side, as the floats of one side will naturally lay themselves over the binders of the other. The number of ends in a repeat of the two weaves must either be alike or one a multiple of the other.

Warp Effects

Levantine on 8 shafts straight draw.

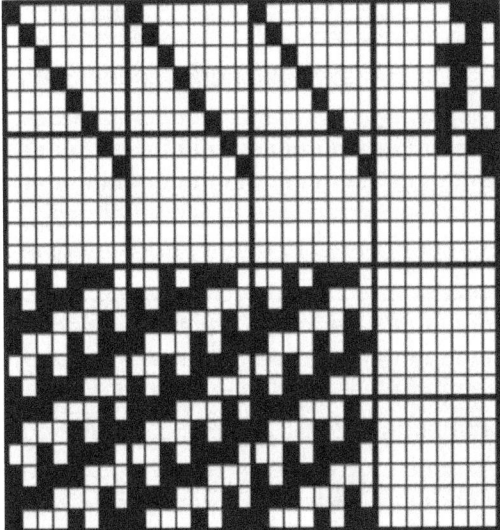

Fig. 74
[pg 56]

Serge 6-2 on 2 sections of 8 shafts each.

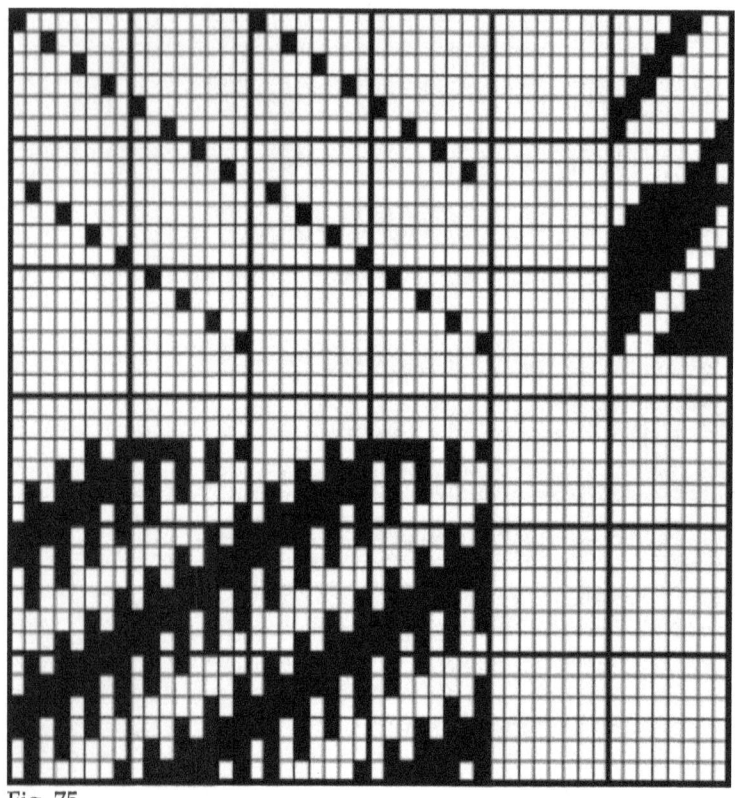

Fig. 75

8 *shaft satin* on 2 sections of 8 harness each.

Fig. 76
[pg 57]

12 *shaft satin* on 2 sections of 12 shafts each.

Fig. 77

Cannele of 8 picks on 2 sections of 4 shafts each.

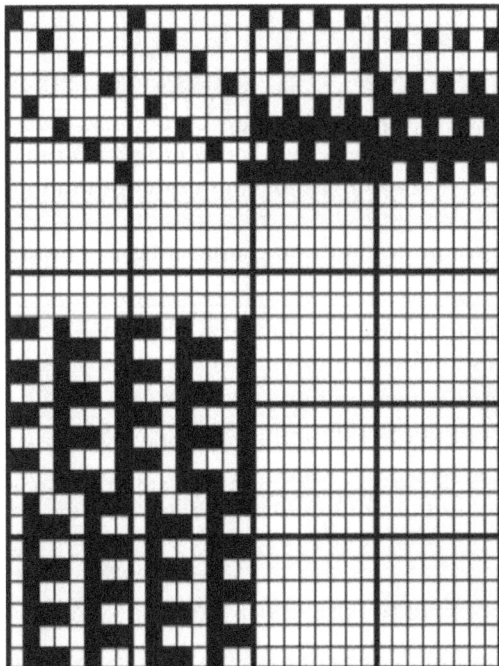

Fig. 78
[pg 58]

FILLING EFFECTS

Serge 5-1 on 6 harness straight draw.

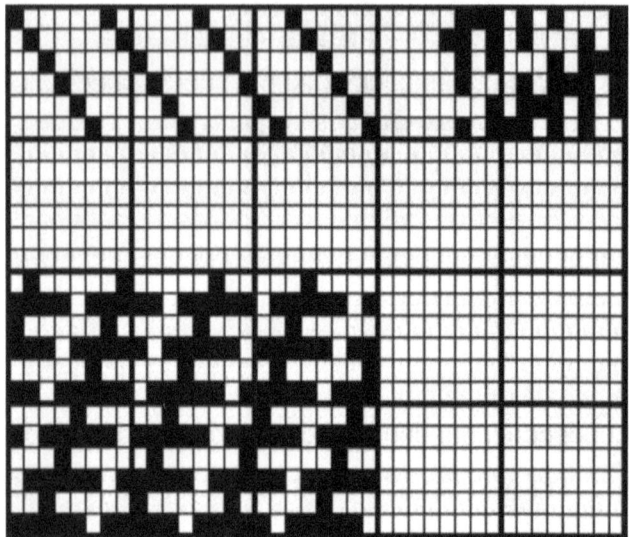

Fig. 79

10 *harness satin* on 10 shafts skip draw.

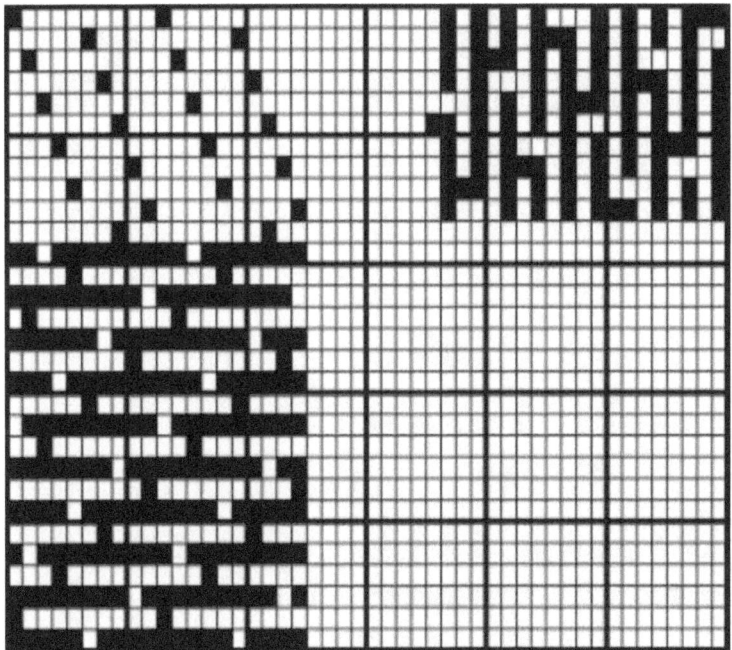

Fig. 80
[pg 59]

Repp on 2 sections of 4 shafts each, 8 threads per section.

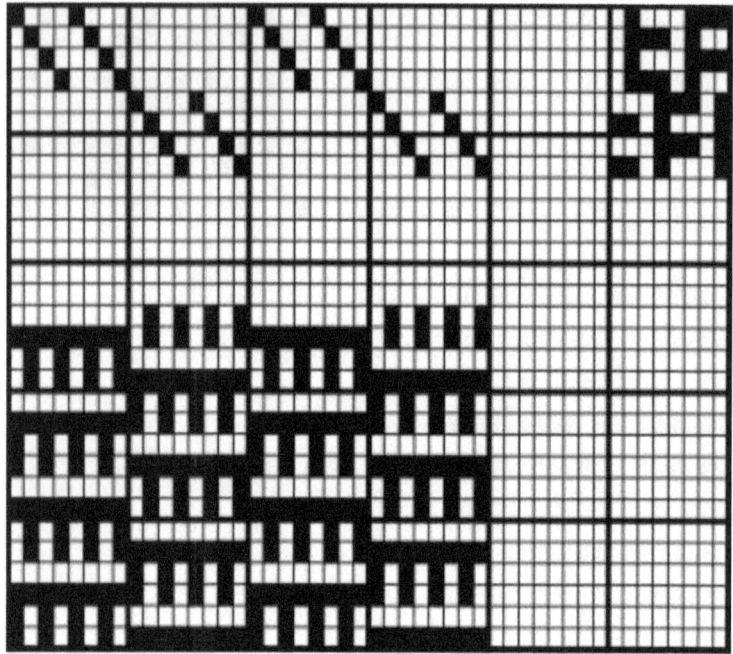

Fig. 81
[pg 60]

PEKINS

With this name we designate fabrics in which stripes of a different interlacing run in the direction of the warp. In combining these weaves it is advantageous to have them contrast distinctly, for instance, a short weave such as taffeta or Gros de Tours, with a longer and looser one such as satin, sergé or cannele, also changes from warp to filling effects. Care must be taken to arrange the joining of the two weaves so that the last thread of one weave will cross the first thread of the other. This will prevent the threads from either stripe to slide over into the other, and so make a clean cut line.

Pekin. A stripe of

2	dents of	8 ends each,	8 shaft Satin,	on 8 shafts straight draw.
12 "		2 "	Taffeta	" 4 " " "

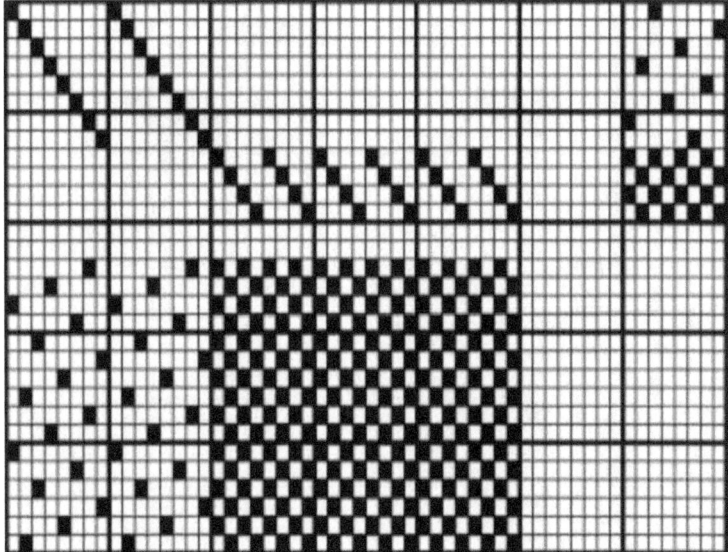

Fig. 82
[pg 61]

Pekin. A stripe of

12 ends Cannele of 6 picks on 1st section of 4 shafts.
12 "   Repp   " 6 threads on 2d and 3d section of 2 shafts each.

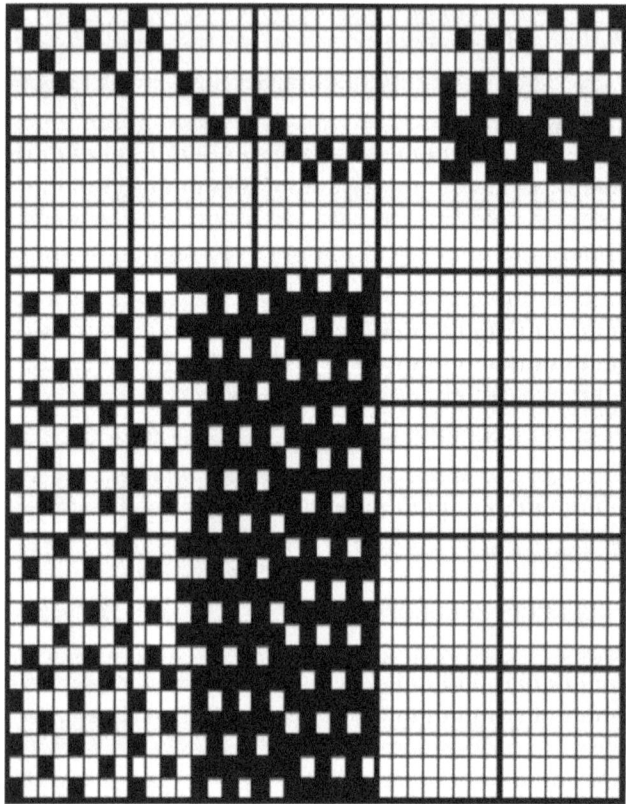

Fig. 83
[pg 62]

Pekin. A stripe of

10 ends	5 leaf Satin	on the	1st section of	5 shafts	straight	draw.
4 "	Taffeta	"	3d "	2 "	"	"
18 "	Serge 3-1, 1-1	"	2d "	6 "	"	"

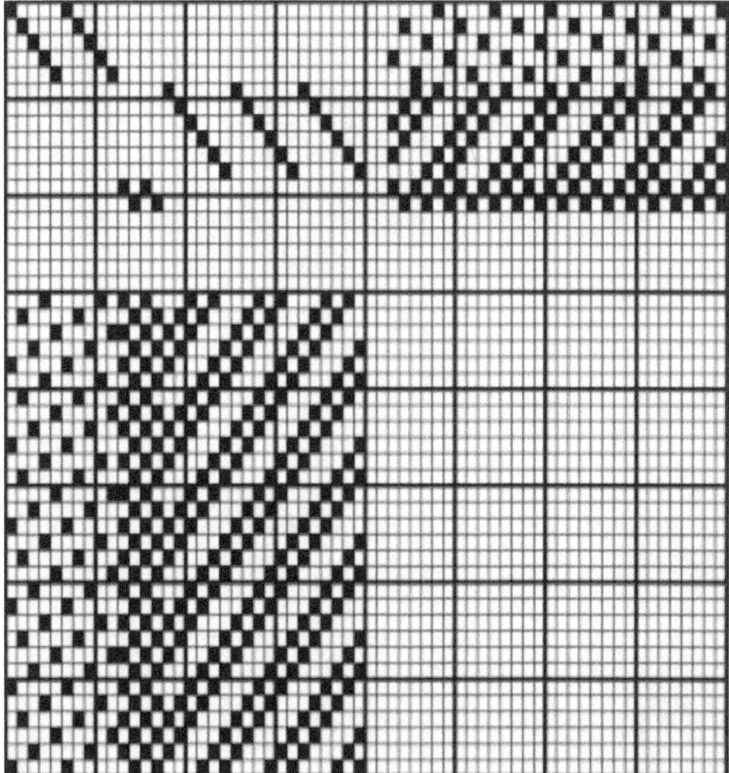

Fig. 84
[pg 63]

BAYADÈRES

While pekins are formed by warp stripes, bayadère shows us stripes of different weaves running in the direction of the filling. The rules given in the previous chapter as to the joining of the weaves will also apply here. The warp which was raised on the last pick of the weave must stay down wherever possible on the first pick of the following weave. The number of shafts employed must go up evenly in the repeat of each one of the weaves that go into the make up of the bayadère.

Bayadère

A stripe of 24 picks Gros de Tours
" 8 " 8-shaft Satin, } on 8 shafts straight through.

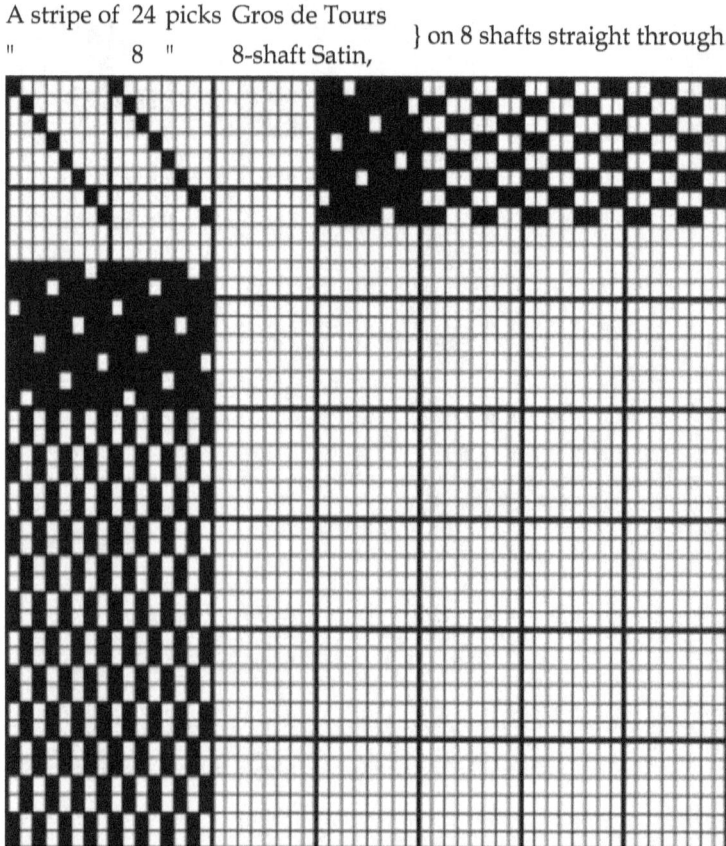

Fig. 85
[pg 64]

Bayadère

A stripe of 18 picks 6-shaft Satin }
" 6 " Serge 5-1, } on 6 shafts, straight draw.
" 4 " Taffeta, }

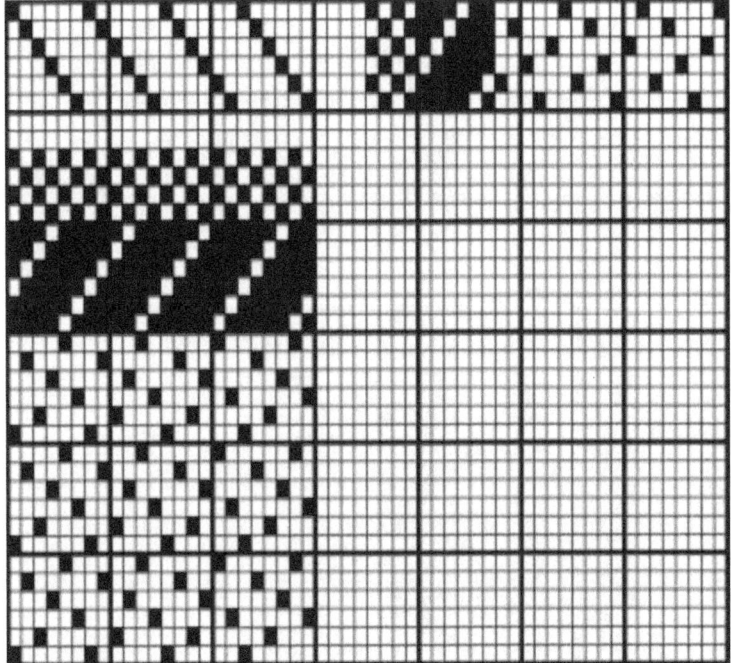

Fig. 86
[pg 65]

CHECKS AND PLAIDS

If pekin and bayadère stripes are combined, we obtain checked fabrics, and of these an endless variety and pleasing effects can be produced with the aid of suitable color combinations.

Check

of 16 threads and 12 picks of the 4 end broken twill,
and 16 "   "   12 "   " Royale of 8 threads,
drawn on 4 sections of 4 shafts each.

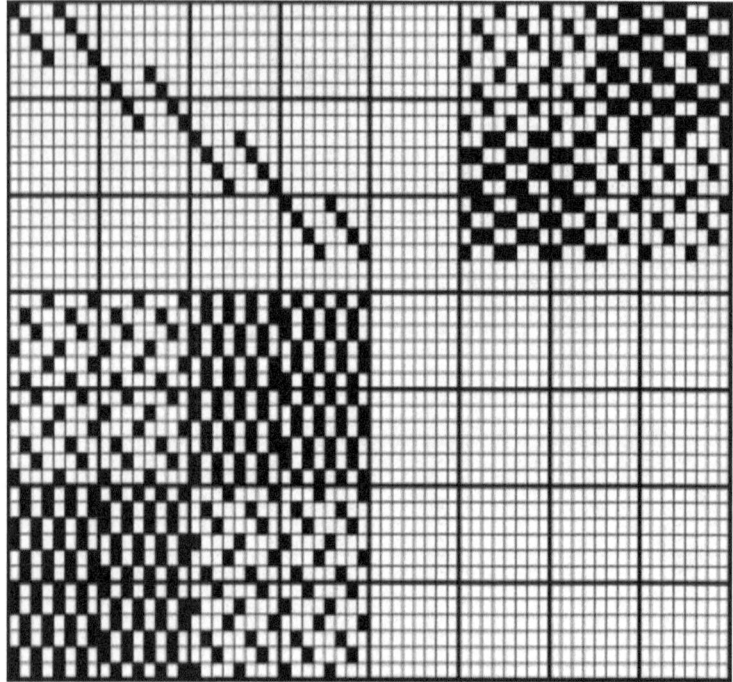

Fig. 87
[pg 66]

Check

of	20	threads	and	16	picks	of	8-shaft Satin,
"	20	"	"	16	"	"	Taffeta,
				4	"	"	Surah 3-1,

drawn on 2 sections of 8 shafts each.

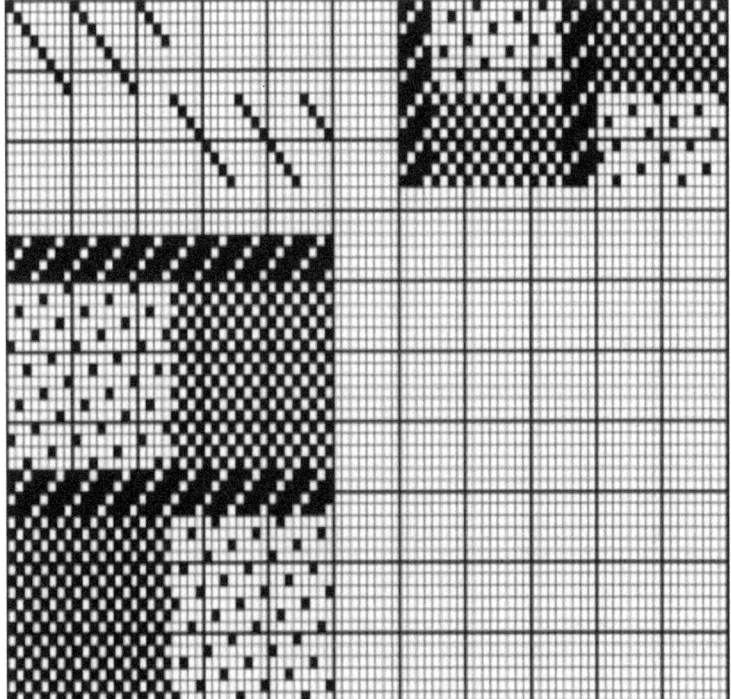

Fig. 88
[pg 67]

DECOMPOSITION

Under the name of *Disposition* we comprise all those points and details which we must ascertain before we can proceed with the construction of a fabric. They are:

1. The weave.

2. Length and width of the cloth.

3. The stock and the dyeing thereof.

4. Reed calculation (number of dents and ends per inch and total number of dents required).

5. Drawing-in the warp in harness and reed.

6. **Texture and arrangement of warp** (warping ticket)

7. **Arrangement of harness.**

8. **Reduction of filling.**

9. **Calculation of warp and filling.**

10. **Finishing.**

11. **Calculation of cost.**

[pg 68]

1. THE WEAVE

If a given sample is to be reproduced or imitated, it must be analyzed, and the following hints will greatly facilitate this operation to the beginner.

Cut the sample straight on two sides, and draw out a number of warp and filling threads until there is a small fringe of perhaps ¼ of an inch. This will allow a thread to be raised a little for examination, without danger of its falling out of the fabric. In most cases it is easier to dissect the filling side, that is, the interlacing of each warp-thread in the threads of the filling system. With the help of the microscope or counting glass we can easily determine over and under how many picks that thread passes and the points of interlacing are accordingly marked on designing paper. This being done for at least the length of a repeat warpways, we take it out and examine the following thread, and so on until the repeat filling-ways is complete. It is advisable to ascertain how many picks ahead or back of the first examined thread the next following one binds before taking the former out.

A knowledge of the construction of weaves as explained in the foregoing chapters will enable us to determine the weave of a simple pattern by merely raising a warp-thread with a needle on any point of interlacing and counting off how many picks apart from this place it makes an impression.

2. LENGTH AND WIDTH OF THE FABRIC

To the length and width which the finished product is to have, we must add a certain allowance for shrinkage and *taking up* of warp during weaving. It may differ from one to ten per cent., according to the texture and weave of the fabric, and can be ascertained with sufficient accuracy by stretching out and measuring a thread of warp and filling and comparing their length with the respective measurements of the sample to be reproduced.

[pg 69]

3. THE STOCK AND ITS DYE

Under this heading we must give the particulars as to nature, *twist*, *quality* and *size* of the silk, and the directions for the dyeing, whether *bright* or *souple*, and in what colors, also whether to be weighted or not.

The size is generally ascertained (in practical work) by comparing it with other silk of which the exact count is known. Another method is to count the number of cocoon threads which a thread of the sample contains, adding to that 1/3 or ¼, according to the quality of the silk; the result will be the count in deniers. To obtain it in drams, divide the latter number by 17,3 as 1 dram is equal to 17,3 deniers.

Suppose we find 20 cocoon ends in a thread of silk: add 1/3, and we have 26-2/3 deniers, which, divided by 17,3 make 1,54 drams.

As silk is always more or less uneven, it is safer to count the cocoon fibers of several threads and to take average thereof.

It requires the experience of years to judge with any degree of certainty as to the origin and quality of silk, whether it be "classical," "extra," "sublime," etc.

There are machines wherewith to ascertain exactly the twist, that is, the number of turns the silk has received in the throwing process.

In the dyeing we distinguish two great classes, of which the names themselves give a good definition. "Bright" has a brilliant luster, while "souple" has more of a dull, subdued appearance. To find out whether the silk has been weighted in the dyeing process, we may compare it with other silk of which the exact conditions are known, or we may burn a small quantity of it. Unweighted silk does not burn readily and leaves a residue of white ashes, while heavy weighted silk burns lively, leaving black, charry ashes.

[pg 70]

4. REED CALCULATION

We count the number of repeats of the weave in a given space, generally ¼ or ½ inch, and multiply this with the number of threads one repeat contains, which gives us the reduction of the warp.

Suppose we had a taffeta, which, as we know, has only 2 ends to a repeat, and counted 30 interlacings per ¼ inch on one pick; we would have 60 threads per ¼ inch or 240 per one inch. In this case the reed may be 80 by 3 or 60 by 4.

Another instance: In an 8-shaft satin we count 10 warp-threads, which bind on the same pick in ¼ inch; this, multiplied by 8, equals 80 ends per ¼, or 320 per one inch; the reed will be an 80 with 4 in a dent or a 64 by 5.

In short, the number of the reed is found by dividing the number of warp-threads that are to go in one dent, in the number of ends per inch. Sometimes, the reed marks are clearly visible in a sample by holding the latter against the light. Silk fabrics move with very few exceptions within the limits of 50 and 90 dents per inch.

To learn the full number of dents required for the width of the cloth, simply multiply the dents per one inch with the width, adding a certain allowance for shrinkage. The edges, of course, must also be taken in consideration, and very often the dents that are taken up by the latter are used to counter-balance that shrinkage.

5. DRAWING-IN THE WARP IN HARNESS AND REED

Here we must specify the number of dents that contain the same number of ends, and whether the latter are single or double, also the number of shafts and the method of drawing-in.

[pg 71]

6. WARPING TICKET

To make out the warping ticket, we need to ascertain the total number of ends, whether leased single or double, and the arrangement of the colors.

7. HARNESS ARRANGEMENT

This is governed by the number of ends to be drawn in and the necessary shafts. If we have, for instance, 100 threads per inch to be drawn on 4 shafts, we must give each shaft 25 heddles per inch. There are generally between 25 and 45 heddles per inch on one shaft.

8. REDUCTION OF THE FILLING

Here we state the number of picks per inch, give directions as to doubling, if such is necessary, and if more than one color or shuttle is used, the rotation thereof.

9. CALCULATION OF WARP AND FILLING

The system adopted in this country for specifying the size of silk is based on the weight in drams (avoirdupois) of a skein containing 1000 yards. A skein, thus weighing 5 drams, is technically called 5-dram silk. The number of yards of 1-dram silk to a pound must accordingly be 256000. The formulas for figuring the amount of silk required for a piece of cloth are as follows:

Warp Calculation

Multiply: $\dfrac{\text{Number of ends} \times \text{length} \times \text{count}}{1000 \text{ yards} \times 256 \text{ drams}} = \text{lbs}$
Divide by:

[pg 72]

Filling Calculation

Multiply: $\dfrac{\text{Picks} \times \text{xply} \times \text{width} \times \text{length of piece} \times \text{count}}{1000 \text{ yards} \times 256 \text{ drams}} = \text{lbs}$
Divide by:

The result in both cases will be in pounds.

The system of grading the silk which is in vogue in Europe, and which is employed by a number of mills on this side, is as follows:

1 skein of 500 meters, weighing 0,05 grams = 1 denier international

or 1 " 476 " " 0,053 " = 1 " Turin system

or 1 " 476 " " 0,051 " = 1 " Milan "

The warp calculation, taking the international denier, would run:

$$\frac{\text{ends in warp} \times \text{length} \times \text{denier} \times 0{,}05 \text{ gram}}{500 \text{ meters}}$$

divided by: 500 meters

for the filling: $$\frac{\text{Picks per meter} \times \text{xply} \times \text{width} \times \text{denier} \times 0{,}05 \text{ gram}}{500 \text{ meters}}$$

divided by: 500 meters

Result in metric weight, kilograms and grams.

10. FINISHING

Give directions as to the process of finishing to which the goods are to be subjected, whether to be pressed, calendered, sized, moiréd, etc.

11. CALCULATION OF COST

If all the foregoing conditions are ascertained, and a sample or a piece of the fabric executed, it remains to the manufacturer to determine the exact figure at which he can produce the article. That this must be done with great accuracy is naturally of the utmost importance, and the calculator [pg 73] must know in the first place the raw stock prices, and also be acquainted with the details of the manufacturing process and the rates of wages paid therein. As a rule, the manufacturer establishes a scale of prices covering all the items of labor cost, mill expenses, etc., and uses this as a basis for his calculations.

A rule or formula for this operation cannot very well be given, as the methods vary in almost every establishment, each choosing the one best adapted to its ideas or dictates of circumstances and conditions.

[pg 75]

DISPOSITION 1

Taffeta glacé

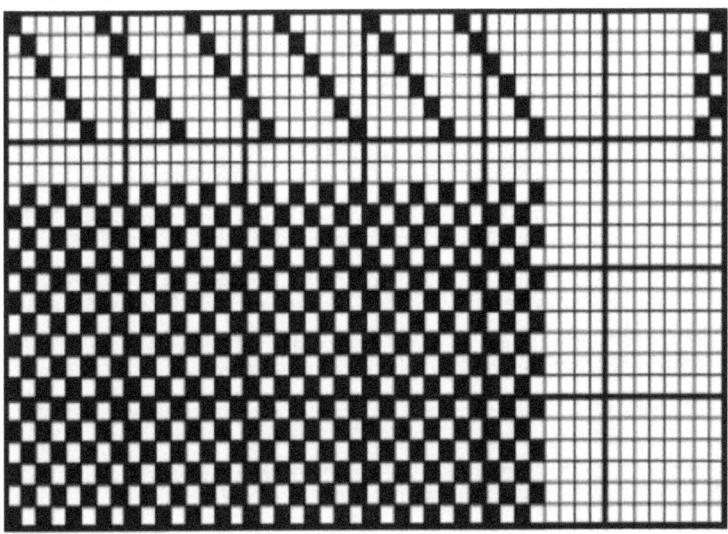

Fig. 89

[pg 76]

Length and Width — One piece 100 yards long, 18¾ inches wide.

Stock and Dye —	Ital. Organ., 24/26 deniers = 1,47 drams,
Warp. —	brown, bright, 14/16 oz.
Filling. —	Japan Tram, 28/30 deniers = 1,65 drams, gold, bright 14/16 oz.

Reed Calculation — Per 1 inch, 70 dents at 3 single ends.

"	18¾ "	1313 "	
	add	13 "	= 1% for shrinkage
		— —	
Total . . .		1326 dents	

Drawing in — 1st edge, 9 dents at 3 double ends.
Ground, 1308 " at 3 single "
2d edge, 9 " at 3 double "
on 6 shafts, straight through.

Warping Ticket — 1st edge, 27 double ends, brown.
Ground, 3924 single " "
2d edge, 27 double " "
— —
Total ... 4032 single ends.

Length of warp 110 yards, including 10% for take up.

Harness — 6 shafts, 3978 heddles per 19 inches.
1 " 35 " " 1 "

Reduction of Filling — Per inch, 100/102 picks, 2 ends.

Warp Calculation — ? lbs = 4032 ends.
1 end = 110 yards.
1000 yards = 1,47 drams.
256 drams = 1 lb.
100 = 104 (4% waste).

$$\frac{4032 \times 110 \times 1{,}47 \times 104}{1000 \times 256 \times 100} = 2{,}65 \text{ lbs., or 2 lbs. 10,4 oz.}$$

[pg 77]

Filling Calculation — ? lbs. = 100 yards
1 yard = 36 inches.
1 inch = 102 picks.
  2 ends.

1 pick = 19 inches.
36 inches = 1 yard
1000 yards = 1,65 drams.
256 drams = 1 lb.
100 = 108 (8% waste).

$$\frac{100 \times 36 \times 102 \times 2 \times 19 \times 165 \times 108}{36 \times 1000 \times 256 \times 100} = 2{,}73 \text{ lbs., or 2 lbs. 11,7 ozs.}$$

[pg 79]

DISPOSITION 2

Surah 3-1

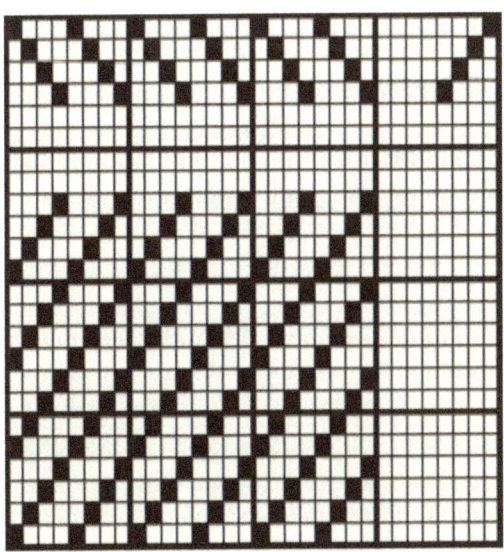

Fig. 90

[pg 80]

Length and Width—1 piece 50 yards long, 23½ in. wide.

Stock and Dyeing— Ital. Organ., 1,30 drams = 20/24 deniers, black,
Warp.— bright, 40/50% = 22/24 ozs.

Filling.— Cotton No. 120/2 black.

Reed Calculation— Per 1 inch, 80 dents at 2 single ends.
" 23½ inch 1880 "
add 36 " = 2% for shrinkage
— —
Total . . . 1916 dents

Drawing in — 1st edge,   8   dents   2 double   ends.
Ground,   1900   "   2 single   "
2d edge,   8   "   2 double   "
on 4 shafts, straight draw.

Warping Ticket — 1st edge,   16   double   ends,   black.
Ground,   3800   single   "   "
2d edge,   16   double   "   "
     — —
Total ...   3864   single   ends.

Warp 54 yards long = 8% for take up.

Harness — 4   shafts,   3832   heddles   per   24   inches.
1   "   40   "   "   1   "

Reduction of Filling — Per 1 inch, 120 picks, 1 end.

Warp Calculation — ? lbs =   4264 ends.
1 end =   54 yards.
1000 yards =   1,30 drams.
256 drams =   1 lb.
100 =   104 (4% waste).

$$\frac{3864 \times 54 \times 1{,}30 \times 104}{1000 \times 256 \times 100} = 1{,}10 \text{ lbs., or } 1 \text{ lb. } 1{,}06 \text{ oz.}$$

[pg 81]

Filling Calculation — ? lbs. =   50 yards
1 yard =   36 inches.
1 inch =   120 picks.
1 pick =   24 inches.
36 inches =   1 yard
840 yards =   1 skein.
(No. 120/2) 60 skeins =   1 lb.
100 =   110 (10% waste).

$$\frac{50 \times 36 \times 120 \times 24 \times 110}{} = 3.14 \text{ lbs., or } 3 \text{ lbs. } 2.24 \text{ ozs.}$$

36×840×60×100
[pg 83]

DISPOSITION 3

Satin Duchesse.

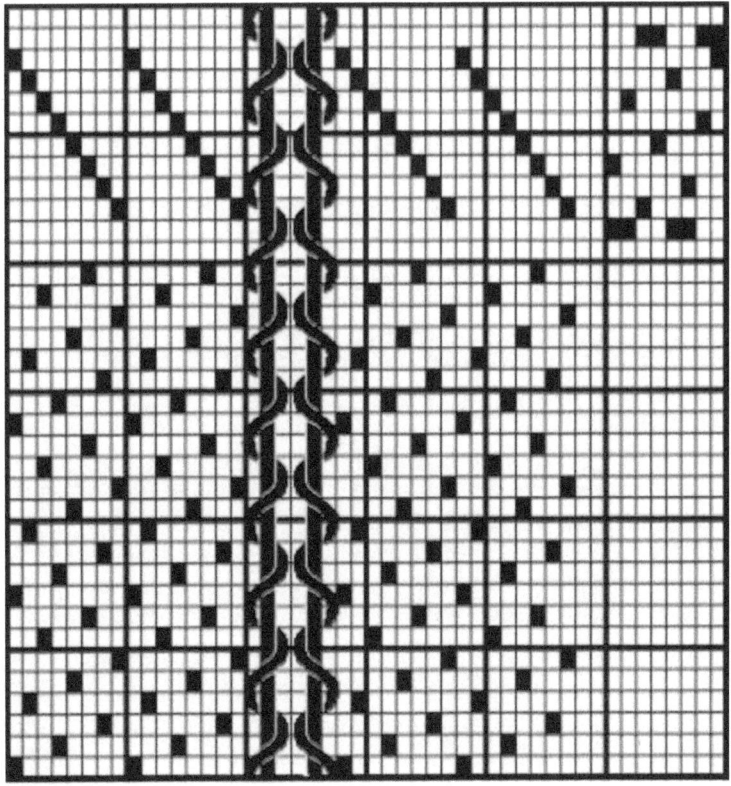

Fig. 91
[pg 84]

Length and Width—One warp 300 yards long, 10 pieces of 60 yards 23 inches wide, 2 pieces in width, with one cut edge.

Stock and Dyeing—
Warp.— Ital. Organ., 1-55/100 drams, black, bright, 20/22 oz.
Filling.— Jap. Tram., 1-8/10 and 2-7/10 drams, black,

souple, 40/44 oz.

We should use for this fabric 1 end filling, 5-thread, 4-5/10 drams, but as we have none of this size on hand, we take: 1 end, 2 thread, 1-8/10 drams, and 1 end, 3 thread, 2-7/10 drams.

Reed Calculation—

	1	inch,	66	dents	4 single ends.
	46	"	3036	"	
		add	54	"	(1¾% shrinkage).
			— —		
Total . . .			3090	dents	

Drawing in—

	1st edge,	{ 2	dents	6×2,	black.
		{ 13	"	4×2,	white.
	Ground,	1514	"	4×1,	black.
		{ 2	"	6×2	black.
	Cut Edge	{ 1	"	2×4,	"
		{ 3	"	empty.	
		{ 1	"	2×4,	"
	Ground,	1514	"	4×1,	black.
	2d edge,	{ 2	"	6×2,	black.
		{ 13	"	4×2,	white.

Satin on 8 shafts, straight draw.
Cross-thread for split edge on 3 shafts, see design.

Warping Ticket—

	Edge,	{ 12/2	black.	}
		{ 52/2	white.	}
	Ground,	6056/1	black.	} twice over.
	Edge,	{ 12/2	black.	}
		{ 52/2	white.	}

12208	ends	black,	{ 318 yards long.	
416	"	white,	{ = 6% shrinkage.	

2/4 black for ground thread, 300 yards long.
2/4 " " whip " 360 " "
each one on a separate little roll.

[pg 85]

Harness—8 shafts, 12368 heddles 47 in.
1 " 33 " 1 "

for the split edge 3 shafts, of which one has only half a heddle.

Filling—Per 1 in., 80 picks, 2 ends (as described before).

Warp Calculation—? lbs =	12224 ends, black (416 white).
1 end =	318 yards.
1000 yards =	1,55 drams.
256 drams =	1 lb.
100 =	104 (4% waste).
(416) $\dfrac{12224 \times 318 \times 1.55 \times 104}{1000 \times 256 \times 100}$	= 24,477 lbs. org. black. = 0,833 " " white.

Filling Calculation—? lbs. =	300 yards
1 yard =	36 inches.
1 inch =	80 picks.
1 pick =	47 inches.
36 inches =	1 yard
1000 yards =	4,5 drams.
256 drams =	1 lb.
100 =	108 (8% waste).
$\dfrac{300 \times 36 \times 80 \times 47 \times 4,5 \times 108}{36 \times 1000 \times 256 \times 100}$	= 21,414 lbs.

[pg 87]

DISPOSITION 4

Armure Satinée

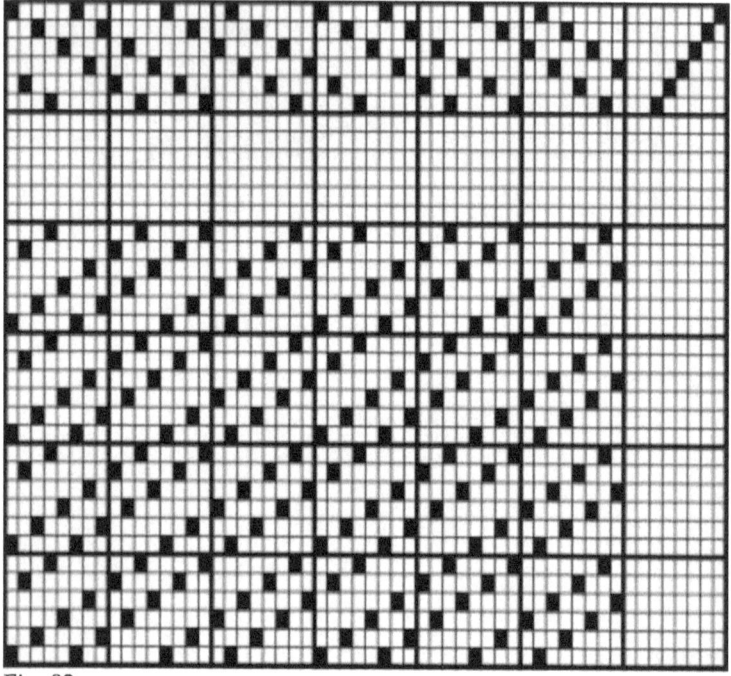

Fig. 92
[pg 88]

Length and Width—One piece 60 yards long, 19 in. wide.

Stock and Dye—
Warp.— Jap. Organ., 1,48 drams, black, bright, 18/20 oz.

Filling.— 3 threads Jap. Tram., 2,95 drams, black, bright, 24/26 oz.

Reed Calculation—
Per 1 inch, 60 dents × 4 single ends.
" 19 " 1140 "
32 " (3% for shrinkage).
— —
Total ... 1172 dents

Drawing in—1st edge, 12 dents × 4 double.
Ground, 1148 " × 4 single.

2d edge,	12 "	× 4 double.	

on 12 shafts, straight through,
or on 6 shafts, as design indicates.

Warping Ticket—1st edge, 48/2 black.
Ground, 4592/1 "
2d edge, 48/2 "

— —

Total . . . 4784 single ends 64 yards long.
=6½% for take up.

Harness—6 shafts, 4688 heddles per 19½ inches.
1 " 40 " " 1 "

Reduction of Filling—Per 1 inch, 104 picks, 2 ends.

Warp Calculation—? lbs = 4784 ends.
1 end = 64 yards.
1000 yards = 1,48 drams.
256 drams = 1 lb.
100 = 104 (4% waste).

$$\frac{4784 \times 64 \times 1{,}48 \times 104}{1000 \times 256 \times 100} = 1{,}84 \text{ lbs., or } 1 \text{ lb. } 13{,}44 \text{ ozs.}$$

[pg 89]

Filling Calculation—? lbs. = 60 yards
1 yard = 36 inches.
1 inch = 104 picks.
2 ends.
1 pick = 19½ inches.
36 inches = 1 yard
1000 yards = 2,95 drams.
256 drams = 1 lb.
100 = 108 (8% waste).

$$\underline{60 \times 36 \times 104 \times 2 \times 19{,}5 \times 2{,}95 \times 108} = 3{,}03 \text{ lbs., or } 3 \text{ lbs. } 0{,}48 \text{ ozs.}$$

36×1000×256×100
[pg 91]

DISPOSITION 5

Surface Printed Armure.

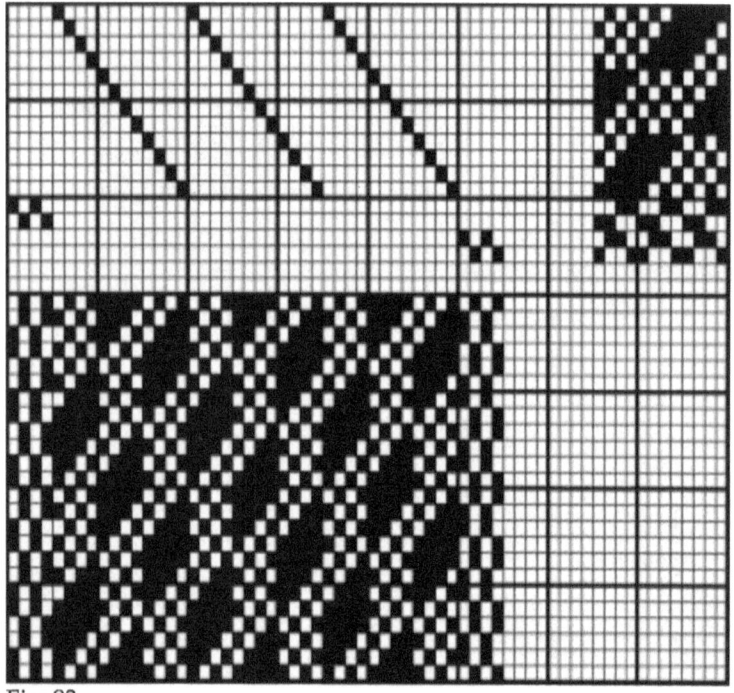

Fig. 93
[pg 92]

Length and Width—One piece 50 yards long, 18½ in. wide.

Stock and Dyeing—
Warp.— Ital. Organ., 1,50 drams, 24/28 deniers, white bright, pure dye.

Filling.— 5 thread Ital. Tram., 4 drams, 13/14 deniers, white, bright, pure dye.

Reed Calculation— Per 1 inch, 52 dents 2×2.
" 18½ " 962 "

	20 "	(2% shrinkage).
	— —	
Total . . .	982 dents.	
Drawing in — 1st edge,	7 dents	4×2 gros de Tours.
Ground,	968 "	2×2 armure
2d edge,	7 "	4×2 gros de Tours.

Armure ground on 12 shafts, straight draw.

Gros de Tours, edges on 4 shafts.

Warping Ticket —	Ground warp, 54 yards. 1936/2 white.
	Gros de Tours edges, 55 yards. 2×28/2 white, on separate rolls.
Total	. . . 3984 single ends.

Harness — 12	shafts,	1936	heddles	per	18½	inches.
1	"	8-2/3	"		1	"

4 Shafts gros de Tours edges, with 28 heddles on each side.

Reduction of Filling — Per 1 inch, 96 picks, 1 end.

Warp Calculation — ? lbs =	3984 ends.
1 end =	54 yards.
1000 yards =	1,50 drams.
256 drams =	1 lb.
100 =	104 (4% waste).

$$\frac{3984 \times 54 \times 1{,}50 \times 104}{1000 \times 256 \times 100} = 1{,}31 \text{ lbs., or 1 lb. } 4{,}96 \text{ ozs.}$$

[pg 93]

Filling Calculation — ? lbs. =	50 yards
1 yard =	36 inches.
1 inch =	96 picks.
1 pick =	19 inches.
36 inches =	1 yard
1000 yards =	4 drams.

256 drams = 1 lb.

100 = 108 (8% waste).

$$\frac{50 \times 36 \times 96 \times 19 \times 4 \times 108}{36 \times 1000 \times 256 \times 100} = 1{,}54 \text{ lbs., or } 1 \text{ lb. } 8{,}64 \text{ ozs.}$$

After weaving, the small flower effects have to be printed on the cloth.

[pg 95]

DISPOSITION 6

Pekin: 8 shaft satin and repp.

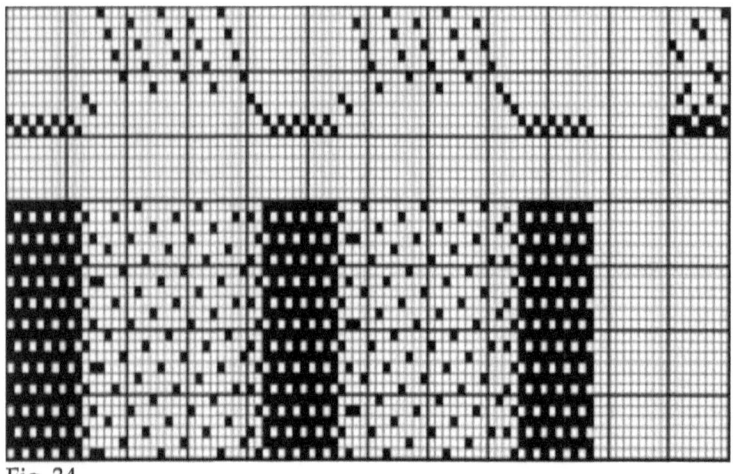

Fig. 34
[pg 96]

Length and Width — 1 piece 100 yards long, 20¼ in. wide.

Stock and Dyeing —
Warp. {
- Ital. Organ., 1-35/100 drams, black, bright, 20/22 oz.
- Ital. Organ., 1-15/100 drams, sky, bright, pure dye
- Cotton No. 100/2, scarlet.

Filling. — 3 threads, Jap. Tram., 2-8/10 drams, scarlet, bright, 14/16 oz.

Reed Calculation— Per 1 inch, 68 dents { 6×1, black satin
{ 6×2, colored "
{ 2×1, " repp.

" 20¼ " 1377 "

add 23 " (1½% for shrinkage).

— —

Total . . . 1400 dents

Drawing in— 1st edge, 7 dents 6×2 satin. }
Ground, 34 " 6×1 " }
2 " 6×2 " }
10 " 6×1 " } twice }
2 " 2×1 repp. } over }
10 " 6×1 satin. }
2 " 6×2 " }
39 " 6×1 " }
5 " 2×1 repp. } twice }
4 " 6×1 satin. } over } 7 times
5 " 2×1 repp. } over
15 " 6×1 satin. }
2 " 2×1 repp. }
15 " 6×1 satin. }
5 " 2×1 repp. } twice }
4 " 6×1 satin. } over }
5 " 2×1 repp. }
5 " 6×1 satin. }
28 " 6×1 " }
2d edge, 7 " 6×2 " }

Satin on 1st section of 8 shaft skip draw.
Binder " 2nd " " 2 "
Repp " 3d " " 2 "

[pg 97]

On each side of every repp stripe two ends of the satin warp must be entered on the 2 binder shafts (2d section), to prevent the ends of the satin to slide over into the repp stripes.

Warping Ticket—I. *Beam satin,* 106 yards.

1st edge,	42/2	black.	
Ground,	204/1	"	}
	12/2	sky.	}
	180/1	black.	}
	12/2	sky.	}
	234/1	black.	}
	48/2	sky.	} 7 times over
	180/1	black.	}
	48/2	sky.	}
	30/1	black.	}
	168/1	"	}
2d edge,	42/2	"	

6132 single ends black.
1680 " " sky.

II. *Beam repp,* 110 yards.

4/1	scarlet.	}
4/1	"	}
10/1	"	}
10/1	"	}
10/1	"	} 7 times over
4/1	"	}
10/1	"	}
10/1	"	}
10/1	"	}

504 single ends scarlet cotton

This warp has to be beamed in stripes. Make out a diagram for the warper the same as shown in Fig. 95.

[pg 98]

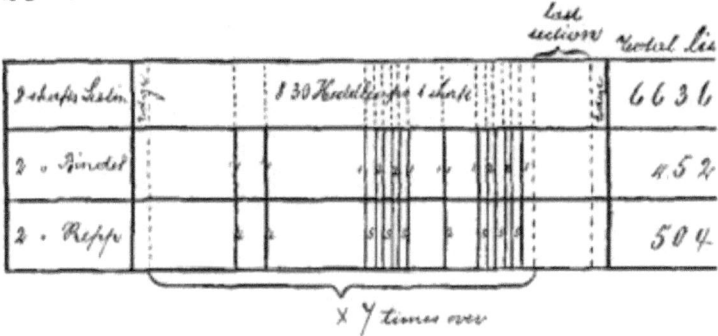

Fig. 95

Harness—Fig. 95 shows how to make a diagram of a harness for a Pekin. The heddles are marked per one shaft. As the repp stripes are only small ones, we use for the satin a full harness, that is, one without open spaces for the repp stripe. All together we have in the satin warp 6888 single and double ends to draw in; of these 252 ends are used for binders, on both sides of the repp stripes. Thus remain for

8 shafts,	6636	heddles	per	20½	inches.
1 "	40	"	"	1	"

Reduction of Filling—Per inch, Per 1 inch, 144 picks, 1 end.

Warp Calculation—? lbs = 6132 ends (1680 ends sky).

1 end = 106 yards.

1000 yards = 1,35 drams (sky 1,15 drams).

256 drams = 1 lb.

100 = 104 (4% waste).

$$\frac{6132 \times 106 \times 1,35 \times 104}{1000 \times 256 \times 100} = 3,56 \text{ lbs. for black, or 3 lbs. 8,96 ozs. 10,4 oz.}$$

$$\frac{1680 \times 106 \times 1,15 \times 104}{1000 \times 256 \times 100} = 0,83 \text{ lbs. for sky, or 0 lbs. 13,28 ozs.}$$

[pg 99]
? lbs = 504 ends.

1 end = 110 yards.
840 yards = 1 skein.
(No. 100/2) 50 skeins = 1 lb.
100 = 104 (4% waste).

$$\frac{504 \times 110 \times 104}{840 \times 50 \times 100} = 1{,}37 \text{ lbs. scarlet cotton, or 1 lb. 5,92 ozs.}$$

Filling Calculation — ? lbs. = 100 yards
1 yard = 36 inches.
1 inch = 144 picks.
1 pick = 21 inches.
36 inches = 1 yard
1000 yards = 2,8 drams.
256 drams = 1 lb.
100 = 108 (8% waste).

$$\frac{100 \times 36 \times 144 \times 21 \times 2{,}8 \times 108}{36 \times 1000 \times 256 \times 100} = 3{,}57 \text{ lbs., or 3 lbs. 9,12 ozs.}$$

[pg 100]

JACQUARD WEAVES

Jacquard weaves usually show on a plain ground figure or flower effects. To obtain these effects the ground is made of one weave, say taffeta, while the figures or flowers are produced in another weave, say satin.

We enter the warp through a jacquard harness, and according to the repeat use a 200, 400, 600, 900, 1200, etc., hook jacquard machine, which means 200, 400, 600, 900 and 1200 ends per repeat.

With a 600-hook machine a larger repeat can be produced than with a 499-hook machine. For instance, if we want to make a cloth the figures of which are to be two inches apart, it is a two-inch repeat, and use the count of 50/4 or 200 ends per inch, we can use a 400-hook machine, 2 in. x 200 ends. But we can make this cloth also on a 1200-hook machine, only the repeat must be designed three times. The advantage of using a 1200-hook machine with the count

50/4 is that 2, 3, 4, 6-inch repeats can be produced, while with a 400-hook machine, only a 2-inch repeat can be made.

There is no end to all the different weaves and the possible flower and other pattern effects that can be made with the jacquard machine. For a jacquard weaving plant the designing is the most important factor.

BOX LOOM WEAVES INCLUDING CREPES

Box looms are required for weaves such as bayadères, checks, plaids (see pages 63-65), bengalines, crêpes, etc., where two or more shuttles are needed to bring out the effect.

For the crêpes (crêpe de chine, crêpe georgette) only two shuttles are needed, while plaids and other articles are made with more shuttles.

To weave such articles the loom or the lay must be fitted with two or more shuttle boxes on one or both sides of the loom.

[pg 101]

If a manufacturer decides to change plain looms to box looms or buy new box looms, it is wise to get 4 × 4 boxes, or four shuttle boxes on each side.

With these looms about everything can be made that is called for in box-loom effects, and as styles change, it is wise to be prepared.

MANUFACTURING COSTS

The calculator first ascertains for the warp: What silk will be used, the cost of the same, total number of ends in the warps for the amount of silk, cost of throwing, dyeing, winding, warping, twisting, entering, and weaving.

Then the filling: Silk to be used, how much, cost of silk, cost of throwing, dyeing, winding, doubling, quilling.

After this determine the cost of weaving, cloth picking, finishing, factory costs, and selling expenses.

To-day most of the operations are paid by "piece work."

The calculator must always take into consideration that poor raw silk or poor dyeing make production slow, increase the cost of labor, and also that more waste will be made.

THROWING: Regular organ usually has 16 turns per inch in the first twist and 14 turns to the inch in the second or reverse twist. Tram receives only one twisting, about three turns to the inch.

As the warp twisting-in is paid for at so much per hundred or thousand ends, no matter how short or long the warp is, it is a saving to make the warps as long as possible, especially in raw, black and staple colors.

The calculator must not forget the cost of entering the first warp in a harness, also the reeding.

Most mills figure the cost of harness and reed in the expense accounts. If plain and fancy goods are made an extra percentage should be figured for the latter.

[pg 102]

EXPENSES are figured differently, as almost every manufacturer has his own system. If a mill makes only a few staple articles it is easy to put down the cost of expenses. Say the mill has a production of 500,000 yards per year, that the expense amounts to $35,000, the cost then is 7 cents per yard.

Manufacturers making all kinds of goods sometimes figure the expenses in percentage, say, for plain goods, with a few picks, like gros-grain, peau de soie, etc., 10 per cent. per yard. Taffeta, satin, etc., having more picks, 12½ per cent. per yard, and fancy and jacquard goods, 15 per cent. per yard. In the expense account we include all charges except raw silk, throwing, dyeing and piece work.

SELLING EXPENSES. Before a calculation is finished we must add the selling expenses to the cost, also take account of the trade discount. Small mills usually sell through a commission house, which pays all expenses and charges a certain commission. Many large firms have their own selling end, and some have their sales guaranteed by a commission house or a bank.

[pg 103]

CALCULATIONS

The prices marked in the following calculations are about as in "normal times." Absolutely correct piece work prices cannot be given as different localities have different prices.

Calculations are usually made per 100 yards, 100-meter warps.

Most goods gain from 3 to 7 per cent. in weaving. That is, if we make a warp of 300 meters for a satin and we obtain 315 yards of cloth, this gain should not be calculated, as usually there is no account taken of samples used in the selling department. But the loss in length should be figured and taken account of on goods with a heavy rib, such as moiré, faille, etc.

DISP. 7 — A 3 1/3-inch repeat can be obtained with a 600-hook jacquard machine, seven repeats in a width of 23 inches.

DISP. 8 — Taffeta weave, but the two cotton picks must go in one hole. This article can only be made with at least two shuttle boxes on each side. For warping use a single and double cross reed, heavy cotton, no knots must be tied.

DISP. 10 — This article must be warped with as much tension as possible and no knots should be tied in. Silk is to be delivered on bobbins from throwster.

CANTON CREPE

DISP. 9 — Can also be made with Canton silk for filling and may be called Canton crêpe. As Canton silk is much cheaper than Japan, the manufacturer can use 4-thread Canton instead of 3-thread Japan for filling at a little difference in cost, thus the cloth will be heavier, but Canton silk is not as even and clean as Japan.

[pg 104]

Article — TAFFETA GLACE Reed 70/3

Disp. 1   Width 18¾ in.

Warp—	Ital. Ex. Class 12/14	$5.00	raw silk
		.60	throwing
	2-Thread Organ. brown bright	.45	dyeing
	16-oz.	.15	winding

	— —	
raw lbs. 2.60	$6.20	$16.23
Warping—4032 at 3¢.		1.21
Twisting—3978 at 25¢. per 300 meters		.33
Filling—Jap. Tram. Best No. 1 13/15	$4.25 raw silk	
	.45 throwing	
2/2 ends 104 picks	.45 dyeing	
	.15 winding	
gold bright 16 oz.	.15 doubling	
	.10 quilling	
	— —	
raw lbs. 2.73	$5.55	15.02
Weaving		7.00
Picking		1.00
Finishing		1.00
Expenses		7.00
		— —
		$48.79

5% trade discount
7½% selling commission
Divide by 87½
Cost per yard = $0.5576

Article—SURAH 3-1 Reed 80/2
Disp. 2 Width 36 in.

Warp— Ital. Ex. Class. 10/12	$5.20	
	.65	
2-Thread Organ. bright black	.81 discount 15%	
24 oz.	.30 dyed 20%	
	— —	
lbs. 3.36	$6.96	$23.39
Warping—5904 at 2¾¢.		1.62

Twisting — 5872 at 35¢. — 600 lb. warp		.25
Filling — cotton 120/1	$1.25	
	.10 dye	
black 1 end 120 picks	.08 winding	
	.07 quilling	
	— —	
lbs. 9.50	$1.50	14.25
Weaving		12.00
Picking		1.50
Finishing		2.00
Expenses		7.00
		— —
		$62.01

5% discount
6% commission
Divide by 89
Cost per yard = $0.6967

[pg 105]
Article — SATIN DUCHESSE Reed 66/4
Disp. 3   Width 2×23 in.

Warp — Ital. Ex. Class. 12/14	$5.00	
	.60	
2-Thread Organ. black bright	.73	
22 oz.	.21	
	— —	
lbs. 8.50	$6.54	$55.59
Warping — 12624 at 2¾¢		3.47
Twisting — 12256 at 25¢. 600 meters		.51
Filling — Jap. Tram No. 1 — 14/16	$4.00	
	.40	
black souple 44 oz.	1.60 net	

1/2 and 1/3 = 1/5 80 picks	.40	
	.40 doubling	
	.30	
	— —	
lbs. 7.14	$7.10	50.69
Weaving		13.00
Picking		2.00
Finishing		2.50
Expenses		8.00
		— —
		$135.76
	Divide by 89	
	Cost per yard = $0.7627	

Article — ARMURE SATIN Reed 60/4
Disp. 4   Width 26 in.
STOCK AND DYE

Warp — Jap. Ex. 12/14	$4.75	
	.60	
2-Thread Organ. black bright	.64	
20 oz.	.19	
	— —	
lbs. 4.30	$6.18	$26.57
Warping — 6520 at 2¾¢.		1.79
Twisting — 6424 at 25¢. 600 meters		.27
Filling — Jap. Tram No. 1 16/18	$4.00	
	.30	
black bright 26 oz.	.94	
	.25	
2/3 ends 104 picks	.25	
	.17	
	— —	

lbs. 6.80	$5.91	40.19
Weaving		10.00
Picking		1.00
Finishing		1.00
Expenses		7.00
		— —
		$87.82
	Divide by 89	
	Cost per yard = $0.9867	

[pg 106]

Article — PRINTED ARMURE Reed 52/2/2

Disp. 5   Width 18½ in.

STOCK AND DYE

Warp —	Ital. Ex. Class. 12/14	$5.00	
	2-Thread Organ. white bright	.60	
	P.D.	.23	
		.12	
		— —	
	lbs. 2.62	$5.95	$15.59
Warping — 3984/1 at 3¢.			1.20
Twisting — 1992/2 at 30¢. 300 meters			.20
Filling — Ital. Tram. souple 13/14		$4.50	
		.30	
	white bright P.D.	.23	
		.10	
	1/5 end 96 picks	.07	
		.17	
		— —	
	lbs. 3.08	$5.20	16.02
Weaving			8.00
Picking			1.00

Finishing	5.00
Expenses	8.00
	— —
	$55.01

Divide by 87½
Cost per yard = $0.6287

[pg 107]
Article — SATIN STRIPED REPS Reed 60/62
Disp. 6   Width 20¼ in.
STOCK AND DYE

Warp —	Ital. Ex. Organ. 24/26	$5.00	
	black bright 22 oz.	.60	
		.73	
		.21	
		— —	
	lbs. 3.56	$6.54	$23.28
	Ital. Ex. Organ. 18/20	$5.20	
		.65	
	sky bright P.D.	.23	
		.12	
		— —	
	lbs. .83	$6.20	5.15
	Cotton 100/2	$1.00	
		.12	
		.08	
		— —	
	scarlet lbs. 1.37	$1.20	1.64
Warping — 8316 at 4¢.			3.32
Twisting — 7434 at 50¢. 300 meters			1.24
	Filling — Jap. tram. No. 1 14/16	$4.00	
		.35	

scarlet bright 16 oz. 1/3 ends	.45	
	.15	
144 picks. lbs. 3.57	.10	
	— —	
	$5.05	18.04
Weaving		12.00
Picking		1.00
Finishing		1.00
		— —
		$66.67
Expenses 15%		10.00
		— —
		$76.67

Divide by 87½
Cost per yard = $0.8763

[pg 108]
Article — MESSALINE BROCADE Reed 60/3
Disp. 7 Width 23 in.

Warp — Jap. Ex. 13/15	$4.50	
	.60	
2-Thread navy bright 16 oz.	.45	
	.15	
	— —	
lbs. 3.20	$5.70	$18.24
Warping — 4320 at 3¢.		1.30
Twisting — 4260 at 30¢. 300 meters		.41
Filling — Jap. Tram No. 1 13/15	$4.00	
	.35	
emerald ex. bright dye 24 oz.	.95	
	.25	
1/3 ends 100 picks	.15	

lbs. 2.50	$5.70	14.25
Weaving		9.00
Picking		1.00
Finishing		1.00
		——
		$45.20
Expenses 15%		6.78
		——
		$51.98

Divide by 87½
Cost per yard = $0.5941

Article — BENGALINE Reed 72/2 by 1 single; 1 double
Disp. 8   Width 36 in.

Warp — Jap. ex. 13/15	$4.50		
	.60		
2-Thread Organ. black bright	.64		
20 oz.	.19		
	——		
lbs. 5.77	$5.93		$34.22
Warping — 7964 at 2¾¢.			2.19
Twisting — 5296 at 30¢. 600 meters			.27
Filling — 56 picks by	$4.50		
5 organ. 2 Cotton	.60		
	.64		
	.19	.45	
	.12	.10	
	——		
40 picks Organ. as warp	$6.05	.08	6.05
lbs. 1		.07	
16 picks black cotton 15/4		——	

lbs. 22.50	$ .70	15.75
Weaving		12.00
Picking		1.50
Finishing		3.00
Expenses		8.00

$82.98

Divide by 91

Cost per yard = $0.9119

[pg 109]

Article — CREPE DE CHINE (Taffeta weave) Reed 60/2/2

Disp. 9 Width 40 in.
44½ in. in reed

Warp — Jap. Ex. Grege 20/22	$4.50		
	.05	winding	
2% waste lbs. 5.60	$4.55		$25.48
Warping — 5340/2 at 2½¢.			2.67
Twisting — 5340/2 at 25¢. 600 meters			.23
Filling — Jap. No. 1 13/15			
4 ends hard twist 60 turns	$4.00		
	1.25	throwing	
84 picks by 2 right, 2 left twist	0.05	quilling	
25% waste and shrinkage	$5.30		
lbs. 6.00			31.80
Weaving			9.00
Finishing and dyeing			5.00
Picking			1.00
Expenses			7.00

			$82.18
			5% discount
			5% selling expense
			Divide by 90
			Cost per yard = $0.9131

Article — CREPE GEORGETTE Reed 50/2
(Taffeta weave) by 1 right, 1 left twist

Disp. 10 Width 40 in. 47 in. in reed

Warp — Ital. Ex. Class 16/18 Raw	$5.20	
	1.60	
	— —	
2-Thread hardtwist 75 turns	$6.80	$36.72
30% shrinkage and waste lbs. 5.40		
Warping — 4640/1 30/2 4760 at 4¢.		1.90
Twisting — 4700 at 50¢. 600 meters		.40
Filling — Same silk as warp		
100 picks by 2 right, 2 left twist	$5.20	
	1.60	
	.10	
	— —	
lbs. 5.00	$6.90	34.50
Weaving		15.00
Picking		1.00
Finishing and dyeing		6.00
		— —
		$95.52
Expenses 12½%		11.94
		— —
		$107.46
		Divide by 87½
		Cost per yard = $1.2281

[pg 110]

CALCULATION BLANK

```
ARTICLE REED ----------------------------------------------------
---------- DISP WIDTH -------------------------------------------
------------------ WARP -----------------------------------------
------------------ | | | | | -------------------------------+---+--
-+---+---+---+-------- | | | | | -------------------------------+-
--+---+---+---+--------- | | | | | ----------------------------
---+---+---+---+---+--------- | | | | | WARPING | | | | --------
--------------------------------------+---+---+---+---+-------- TWIS-
TING | | | | | ---------------------------------------+---+---+---
+---+-------- FILLING | | | | | ------------------------------
----+---+---+---+---+--------- | | | | | ----------------------------
----------+---+---+---+---+-------- | | | | | WEAVING | | | | -
--------------------------------------+---+---+---+---+-------- PI-
CKING | | | | | ---------------------------------------+---+---+--
-+---+--------- FINISHING | | | | | -----------------------------
----------+---+---+---+---+--------- EXPENSES | | | | | ----------
---------------------------------+---+---+---+---+-------- | | | | | | |
| | -------------------------------------------+---+---+---+---+--------
PRICE PER YARD
```

www.ingramcontent.com/pod-product-compliance
Lightning Source LLC
Chambersburg PA
CBHW031431210526
45464CB00005B/2154